William Wales

The Method of Finding the Longitude at Sea

William Wales

The Method of Finding the Longitude at Sea

ISBN/EAN: 9783337327828

Printed in Europe, USA, Canada, Australia, Japan

Cover: Foto ©berggeist007 / pixelio.de

More available books at **www.hansebooks.com**

THE

METHOD OF FINDING

· THE

LONGITUDE AT SEA,

BY

TIME-KEEPERS:

TO WHICH ARE ADDED,

TABLES OF EQUATIONS

TO

EQUAL ALTITUDES.

More extensive and accurate than any hitherto published.

By WILLIAM WALES, F.R.S.

And Master of the Royal Mathematical School in Christ's Hospital.

LONDON:

Printed by C. Buckton, Great Pulteney Street, Golden Square, for the Author; and fold by him in Christ's Hospital: also by F. Wingrave in the Strand, J. Johnson in St. Paul's Church Yard, W. Richardson in Cornhill, and all other Booksellers.

1794.

PREFACE.

IN the following tract I have shewn, in as plain and clear a manner as I am able;

First, What longitude is.

Secondly, What is understood by the term, Finding it: that is, in what manner, and from what circumstances the knowledge of it is derived.

And, thirdly, The method of applying a watch, or time-keeper, to this purpose; which naturally includes the means and method of finding the rate which the time-keeper goes at, and how much it is too fast or too slow for mean time at the place the ship sails from before the voyage commences. This, indeed, is the only difficult part of the business: for the method of finding the longitude by a time-keeper, when the time-keeper is regulated to keep mean time, and set to mean time at the first meridian; or when its daily deviation from mean time at the first meridian is given, is too plain to be misunderstood, and too simple an operation to require rules for putting it in practice. Even the business of finding the *rate*, as it is called, has so little of what is really difficult in it, that some of the officers in the East-India Company's service constantly found the rates of their time-keepers themselves, without any thing having been written expressly concerning it: nor should I have thought it necessary to write on the subject now, if some very extraordinary opinions had not been lately advanced relative to it. The opinions here hinted at are, indeed, so very extraordinary, and at the same time so void of foundation, that they would neither have deserved, nor obtained notice, if they had not received countenance, and even support, from authority, in itself the most respectable, though, in what relates to science, not infallible. By what means, or for what purpose they obtained such countenance and support, it is neither my business or intention to enquire : but, meeting with it, they have certainly done more than could even have been expected toward bringing time-keepers into disrepute, and toward defeating the endeavours of the Board of Longitude, who have been labouring incessantly for the last thirty years to establish the use of them, and to encourage improvements in their construction. Indeed, every one who has to make use of them, must perceive immediately, that if the new doctrine be true, time-keepers muſt be of very little uſe at ſea, as a means of diſcovering the longitude : and it was the experience, alone, which ſeamen had already had of their utility, in direct oppoſition to theſe opinions, that has pre-ſerved their reputation.

In proof of what is here advanced, I can aſſure the reader, that I have heard more doubts expreſſed, and more difficul-ties ſtarted relative to what is called the rate of a time-keeper's going, the proper mode of obtaining it, and the degree of dependence which may be placed on them, in the laſt twelve or eighteen months, than in all my life before, though I have been pretty intimate with the ſubject for near forty years : and I am firmly perſuaded that if the uſe of time-keepers had not been well eſtabliſhed by experience before theſe new doctrines were broached, they would not have been ſpeedily introduced into practice now.

To remove the ſtigma which, for pri-vate purpoſes, has been unjuſtly thrown on theſe valuable machines, by thoſe who, notwithſtanding, would be thought, and indeed ought, to have acted differently by them; to ſave myſelf, if poſſible, ſome trou-ble in future ; and to enable thoſe whoſe bufinefs it is to apply them to the purpofe of finding the longitude, in the beft, and eafieft manner, were my motives for writing the following tract. But as my firft object may be thought by fome to be of a controverfial nature, I fhall confine what I have to fay on it to this place, and to as few words as poffible : it is comprifed under the four following heads :

1. It has been reprefented, that to render the ufe of a time-keeper reafonably certain, the rate of its going ought to be taken from a period of time, at leaft, equal to that of the voyage which it is intended to be ufed in *.

If this were true, it would be neceffary for the Captains, and Officers, of Eaft-India fhips, to begin to examine the rates of their time-keepers, or to employ people to do it for them, at leaft four or five, or indeed fix months before the time they are appointed to fail, becaufe the outward bound voyage may generally be expected to laft fo long : but this is often four, five, or fix times as long as they know of their appointments before the time when they are directed to have every thing on board, and be ready to fail. On board fhips of war, which are frequently ordered out of port

---

* See Mr. Mudge's " Narrative of Facts," p. 22, &c.

on a few days notice, and remain out for months together, the objection is still stronger. And in voyages of Difcovery, where time-keepers are more efpecially neceffary, and in which fhips are feldom in a port more than a fortnight or three weeks at a time, and at fea four or five months, they would, according to this doctrine, be totally ufelefs. So abfurd are affertions, fometimes, when they are made to ferve particular purpofes; becaufe few of the confequences which may be drawn from them, except that which they are intended to ferve, are confidered.

2. It has been reprefented as a matter of great difficulty to affign the rate which a time-keeper may be expected to keep in future *. It has even been afferted that " no general rule has been laid down, nor feems capable of being found, for affigning, from any number of trials, the rate which the watch fhall be moft likely to obferve in any future trial;" and that the common way, of dividing the whole that a watch has done, during a given period, by the number of days, and applying the quotient as a daily rate to be ufed in any period fucceed-

* See the Report of Mr. Mudge's Committee to the Houfe of Commons, p. 7.

ing, is certainly not a method that can pretend to any extraordinary exactnefs *.

It will be fhewn in the following tract, that there is nothing more fimple in itfelf, or more eafily performed, than finding the rate, which a time-keeper goes at: and I affert, without fear of being confuted, that the rate, fo found, is fufficiently exact for the purpofe it is intended for, if the time-keeper be fo. For every one, who is con-verfant with aftronomical obfervations, knows that the time can be always found within a very fmall part of a fecond. But, that I may not under rate it, I will fuppofe that an error of half a fecond may be com-mitted in finding the time, both at the be-ginning and end of the interval which the rate is taken from, and that the obferver is fo unfortunate as to make it different ways, which will create an error of one fecond in the whole interval : this error, which few who are acquainted with the fubject will admit can happen, will only caufe an error of twelve feconds, or three minutes of lon-gitude in a voyage of a whole year, the rate being taken from a month, as is ufual. And I affert, that this is a general rule for finding the rate which a time-keeper is moft likely to go at in future. If, indeed,

---

* Report of Mr. Mudge's Committee, p. 9.

the irregularities in a watch's going be very great, or the period in which its irregularities return be long, or infinite, I grant that this method of deriving a rate is uſeleſs; and ſo will every other method be: it is, indeed, impoſſible to find the ſum of a ſeries the terms of which bear no relation one to another; but the method is only uſeleſs becauſe the watch is ſo, at leaſt for the purpoſe under conſideration ! It is im-poſſible to avoid expreſſing ſome ſurpriſe at finding ſuch an obſervation made.

3. It has been ſaid that it is not well-agreed what the object and purpoſe is which the taking of a rate is intended to anſwer *: and that among perſons likely to be beſt informed on the ſubject, great diverſity of opinion prevails in regard to what ought to be taken for the rate of going of a watch; and that even the ſame perſons fluctuate, at times, between opinions widely different from each other concerning it †.

I am not in fear of being confuted when I aſſert, that every diſintereſted perſon, who underſtands the ſubject, will agree perfectly concerning the purpoſe which taking the rate of a watch is intended to

* Report of Mr. Mudge's Committee, p. 9.
† Ibid. p. 10.

b

answer, notwithstanding he may express himself obscurely, or even incorrectly, in answering directly to . questions which are put to him, without previous notice, on scientific subjects. Questions of such a nature ought never to be put in this manner, but be given in writing; and the answers should be returned in the same manner, after due deliberation, if correct information be expected on the subject. I am in as little danger of being confuted, when I say that every person of the above description, will agree also about the rate which ought to be taken of a time-keeper to go to sea with. But there may be, and undoubtedly is, in the instance under consideration, a material difference between the means which ought to be made use of for trying how far a machine is adequate to a certain purpose, and in applying it to that purpose in the best manner circumstances will admit of, when the machine is known to be, in some degree, imperfect: and if so, a man cannot be said to fluctuate between " opinions widely different," because, on being asked what he thinks the best method of trying how far a time-keeper is deserving of that name he proposes one method; and, on being asked, afterward, what he thinks the best method

* Report of Mr. Mudge's Committee, p. 10.

of applying an imperfect time-keeper to the purpose it was intended for, he proposes another, somewhat different. This, however, is a just statement of the only case which has been brought forward to prove that "the same persons fluctuate at times between opinions widely different from each other:" for, in the first case, the opinion of the person alluded to had been asked concerning the best means of trying the merits of a time-keeper; and in the latter concerning the means of finding a rate which the time-keeper was most likely to conform to in future. The person, unfortunately, did not perceive that the question had been shifted on him: and I am persuaded the gentlemen who examined him were not aware that they had done it; for, if they were, the inference they have drawn would be very uncandid. Not a single instance has been adduced to shew, that "great diversity of opinion prevails, in regard to what ought to be taken for the rate of going of a watch, among persons likely to be best informed on the subject; nor does it appear to me that one can.

4. It has been asserted that when ever a ship touches at a port, the longitude of which is known, a new rate of a time-keeper may be obtained; and that if a per-

fon land at a place where the longitude is not known, though but for a few hours, the longitude may be determined within lefs than a quarter of a degree, and a new rate of the time-keeper found from it *.

The falfehood of thefe affertions, and the ignorance they betray of the fubject they relate to, are too grofs to impofe on any who underftand that fubject; and would not have been taken notice of here, were it not that many perfons have occafion to ufe time-keepers who are not competent to determine how far fuch affertions are true or falfe; and therefore may be mifled by the boldnefs which they are made with.

To prove the futility of the firft of them, namely, that a new rate of going may be obtained for a time-keeper every time a fhip touches at a port where the longitude is known, I need only put the following cafe: Suppofe a fhip to fail from England, and reach the Cape of Good Hope in fixty days, when it is difcovered that the longitude fhewn by the time-keeper is 30 minutes wrong; or that the watch has erred two minutes in time. It will be obvious to every one, that if the time-keeper altered its rate immediately on

* See the Appendix to the Report of Mr. Mudge's Committee, p. 120, 121.

leaving England, and continued to go uniformly afterward, it muſt have altered its rate exactly two ſeconds a-day. If it did not alter its rate till one-third part of the time had elapſed, and went uniformly afterward, it muſt have altered its rate three ſeconds a-day. If it did not alter its rate till half the time was expired, it muſt have altered its rate four ſeconds a-day. If not till two-thirds of the time were elapſed, it muſt have altered its rate ſix ſeconds a-day: and if not till the day before they arrived at the Cape, which is as probable as the firſt ſuppoſition, it muſt have altered its rate of going *two minutes* a-day! If we conceive the watch to have altered its rate by degrees (which is moſt probable;) or if we admit the probability, (which is no ſmall one) that it might firſt go faſter and afterward ſlower; or firſt ſlower and afterward faſter, than it was going at the commencement of the voyage, the quantity by which a rate, ſo obtained, may be uncertain, as well as the number of ways it may happen, cannot be computed. It is eaſy to conceive that ſuch an opinion might have been advanced if a man had been taken by ſurpriſe, and the queſtion had come upon him unexpectedly: but that an aſſertion ſo prepoſterous, could be made premeditatedly, and repeatedly, after the matter had been canvaſſed, to invalidate the teſtimony of others, who had denied the poſſiblity of doing it, would be ſcarcely credible, if it did not ſtand on record under as good authority as any in the kingdom.

In reſpect to the other aſſertion, namely, that the longitude may be found, with certainty to leſs than a quarter of a degree, whenever a perſon can land, though it be but for a few hours; it may be obſerved that correſponding obſervations are out of the queſtion, becauſe the longitude is to be determined on the ſpot, for the purpoſe of finding a new rate for the time-keeper, to be uſed in the future part of the voyage: conſequently, theory muſt be reſorted to for the time under the firſt meridian. This being premiſed, every aſtronomer who knows the preſent ſtate of the lunar theory, and has had much practice in obſerving the diſtance of the moon from the ſun and ſtars, will ſmile at the idea of determining the longitude of a place *with certainty, to leſs than a quarter of a degree*, whenever he can land for *a few hours.* Had the Gentlemen, before whom this was advanced, been inclined to ſift too narrowly into things, might they not have aſked this acute obſerver what occaſion there can be for time-keepers, if this boaſt be true? Or why it was neceſſary to land to do it? Be-

caufe every one who knows any thing of
the matter has long known, that " the
motion of a fhip is no otherwife an impediment in this fort of obfervations, than as
it renders the repetition of them more tedious and troublefome to the obferver *.
And if the longitude of a fhip can be determined with fo much accuracy, and in fo
little time, whenever a man chufes to do
it, furely there can be no occafion for him
to put himfelf to the expence and trouble
of a time-keeper. But we are told further,
that " if they have proper inftruments they
may determine it by Jupiter's Satellites."
In fixed obfervatories, where the obferver
has every advantage and convenience, about
nine or ten of thefe obfervations are, on
a medium, obtained in the courfe of a
whole year; and out of this number, fmall
as it is, it is not unufual for fome of them
to differ from the tables confiderably more
than the quantity here mentioned : and if
this be the cafe in fixed obfervatories,
where they are certain of their time to lefs
than one fecond; what muft the deviation
be when this article is alfo, in fome degree,
uncertain, and the obfervations made under
many difadvantages?

* See Dr. Bradley's Letter to the Board of Admiralty
concerning Admiral Campbell's Obfervations, at the end of
Mayer's Tables.

I have only further to add, that all the neceffary tables, which are not found in the Requifite Tables, are added at the end: and it is to be noted, that when a table is referred to which is in the Requifite Tables, that book is referred to by name; and when the name of that book does not occur, the table, then referred to, will be found at the end of this work.

The tables of equations to equal altitudes are more extenfive than any which have hitherto been publifhed; and I flatter myfelf will be found exceeding accurate. They are general; and will be found ufeful to many perfons befide feamen.

Throughout the whole I have had practice in view; and truft I have, in no inftance, contented myfelf with barely telling my reader he muft do fuch, and fuch things, without explaining the manner in which he is to do them: knowing, from much experience, that inftructions of this kind, and efpecially in what relates to the ufe of aftronomical inftruments, cannot be underftood by any but thofe who are tolerably well acquainted with the fubject before hand: fuch do not want any inftructions, and will, I truft, excufe my prolixity when they know my motives for it.

THE

THE
METHOD

OF FINDING THE LONGITUDE AT SEA,

BY

A TIME-KEEPER.

1. **T**HE Earth is a body compofed partly of land and partly of water, of a globular figure; and it revolves on an imaginary line, paffing through its center, as an axis, once in the courfe of a natural day.

2. The two points where this imaginary axis meets the furface of the earth are called its poles: that which is neareft to Europe being called the *north* pole, and the other is called the *fouth* pole.

3. If a circle be fuppofed to circumfcribe the earth, exactly in the middle between the two poles, that circle is called the Equator: the circle which correfponds to it in the heavens, is called the Equinoctial.

4. The Ecliptic, is that circle in the heavens which the earth defcribes in its annual motion round the fun; being the fame with that which is apparently de-

fcribed by the fun in a year: it cuts the
equinoctial in two points which are directly
oppofite each other. That point where
the ecliptic cuts the equinoctial, and paffes
to the northward of it; is called the vernal
equinoctial point; and the other, where it
cuts the equinoctial, and goes to the fouthward of it, is called the autumnal equinoctial point.

5. The ecliptic is divided into twelve
equal parts, called figns; each fign into
30 degrees, each degree into 60 minutes,
and each minute into 60 feconds. Every
circle is fuppofed to be divided into 360 degrees, a degree into 60 minutes, and each
minute into 60 feconds, and fo on.

6. The fun's longitude is an arch of
the ecliptic, contained between the vernal
equinoctial point and the center of the
fun: it is expreffed in figns, degrees, minutes, and feconds; and is given on p. II.
of the Nautical Almanac for the noon of
every day at Greenwich.

7. Circles defcribed on the furface of
the earth to pafs through both poles, are
called meridians. The circles which cor-
refpond to thefe in the heavens, are called
hour-circles, and circles of right-afcenfion.
A circle of this kind may be defcribed

through any place on the earth, and it is then called the meridian of that place which it is defcribed through.

8. That arch of the meridian which is intercepted between any place and the equator, or the number of degrees, minutes, &c. which are contained in it, is called the latitude of that place.

9. The horizon is that circle which bounds the view of an obferver when he is on the fea, or an open plane.

10. The altitude of the fun, or a ftar, is the number of degrees, minutes, &c. that its center is above the horizon, at any time: when the object is on the meridian of the place, its altitude is the greateft it can have that day; and is called its meridional altitude.

11. The right-afcenfion of the fun, or a ftar, is an arch of the equinoctial contained between the vernal equinoctial point, and that point of it which comes to the meridian at the fame time with the object; or it is the degrees, minutes, &c. contained in that arch. It is fometimes given in time, reckoning one hour for every 15 degrees; and is thus given for the fun, on page II. of the Nautical Almanac every day at noon, under the me-

ridian of Greenwich. The right-afcenfions of fuch ftars as are wanted in this tract are given in Table V.

12. The declination of the fun, or a ftar, is the number of degrees, minutes, &c. which its center is north or fouth of the equinoctial. The declination of the fun is given for the noon of every day at Greenwich on page II. of the Nautical Almanac; and the declinations of fuch ftars as are wanted in this tract will be found in Table V.

13. That part of the equator which is intercepted between the meridian of any place, and the meridian of the place where longitude is fuppofed to begin, or the number of degrees, minutes, &c. contained in it, is called the longitude of that place.

14. Formerly all Geographers reckoned their longitude from the meridian of the ifland of Ferro, the moft wefterly of the Canary Iflands, becaufe it was the moft wefterly land that was known when that practice was adopted; and the longitude was reckoned wholly eaftward, up to 360 degrees. The Dutch, German, and fome other Geographers, ftill reckon their longitude in this manner; but others reckon it from the meridian of the capital city of

their own country, both eaftward and weft-
ward, to 180 degrees; and the Englifh generally reckon it from the meridian of the
Royal Obfervatory at Greenwich, becaufe
all tables are adapted to that meridian, and
all the computations in the Nautical Almanac are made for it.

15. To explain thefe things more par-
ticularly, let E N Q S, fig. 1. reprefent the
earth, P the north pole, the oppofite point
to which, or the fouth pole, falling behind
the figure, is not feen. E A Q will then
reprefent the equator, which being equally
diftant from the two poles, and they 180
degrees afunder, muft evidently be every
where 90 degrees from each of them. PO,
P 15, P 30, &c. are meridians; P O being
that of Greenwich, at which longitude be-
gins according to the Englifh way of
reckoning it, and from which it is always
counted.

16. If the point B be fuppofed to reprefent the ifland of Barbadoes, P B 60
will reprefent the meridian of that ifland,
and the arch B 60 will be its latitude;
and as the ifland lies on the north fide of
the equator, the latitude will be north:
A 60, the arch of the equator intercepted
between the meridians of Greenwich and

Barbadoes, will be its longitude; which is weſt, becauſe Barbadoes lies to the weſt of the meridian of Greenwich.

17. As the earth revolves on its axis from weſt to eaſt once in the courſe of a natural day, or 24 hours, it is manifeſt that the meridian of every place on the earth will come oppoſite to the ſun once in that time : and becauſe that whenever the meridian of any place comes oppoſite to the ſun it is noon at that place, it follows that when the line N P A S, which is the meridian of Greenwich, comes oppoſite to the ſun it will be noon at Greenwich, and a new day will commence. It muſt be farther obſerved, that as the motion of the earth on its axis is uniform, equal parts of the equator will paſs by the ſun in equal ſpaces of time; and, conſequently, that one twenty-fourth part of it, or 15 degrees, will paſs by the ſun in one hour. Hence, at the end of one hour after the time when it was noon at Greenwich, the meridian P 15 C, which is 15 degrees weſt of the meridian of Greenwich, will come oppoſite the ſun, and make noon to all the places which are on that meridian, and a new day will commence at thoſe places exactly one hour after it commenced at Greenwich. In like manner,

two hours after the time when it was noon at Greenwich, the fun will arrive at the meridian P 30 D, which is 30°. weft of the meridian of Greenwich, and make noon to all the places that are on it; and a new day will commence at thofe places, when it is two o'clock in the afternoon at Greenwich : the fame reafoning will hold good for places which are ftill farther weft; and it is therefore evident, that the difference of longitude between any two places, bears the fame proportion to the difference between the times at thefe two places that 15 degrees bear to an hour of time. Confequently, if the difference between the times at two places be turned into degrees, minutes, &c. at the rate of 15 degrees to an hour, it will be the difference of longitude between thofe places : and it is farther manifeft, that if the time at the meridian of Greenwich be greater than the time at the fhip, or other place, the fhip or place lies to the weft of Greenwich; but if the time at the fhip or place be greater than the time at Greenwich, the fhip or place lies to the eaft of Greenwich; becaufe the time of noon at that place muft have preceded the time of noon at Greenwich.

18. Hence it appears, that to find the longitude of any place from another given one, as Greenwich, we muſt find the time of the day at each place, take the difference between theſe two times, and turn it into degrees and minutes by allowing 15 degrees for every hour, or one degree for every 4 minutes of time, and one minute for every four ſeconds of time.

19. The time at the ſhip, or place where the obſerver is ſituated, may be found from an obſervation of the ſun's altitude in the day time, or from the altitude of a known fixed ſtar in the night, by rules which will be given hereafter : for finding the time at the firſt meridian, many methods have been propoſed; but two only have yet been found to perform it with reaſonable accuracy. One is by obſerving the diſtance of the moon from the ſun, or a fixed ſtar; and the other is by means of a Watch, or Time-keeper: it is this latter method, only, which I ſhall conſider at preſent. But before I ſhew how the time is to be found at the firſt meridian by a Time-keeper, it will be neceſſary to explain the difference between mean and apparent time; to ſhew how one may be derived from the other; and how the apparent time may be

found at any place where the obferver is fituated.

20. Apparent Time, called by foreign Aftronomers *true time*, is that which is derived immediately from the fun, either by obferving its tranfit over the meridian, which happens at the inftant of apparent noon, when a new day commences, or by obferving its altitude at a diftance from the meridian.

21. Mean Time is that which is fhewn by good clocks or watches, properly regulated: It is fometimes called equated time, the reafon for which will appear prefently.

22. As the earth revolves uniformly on its axis, if it had no annual motion in its orbit, or if that motion was uniform, and in a plane which is parallel to the plane of the equinoctial, the natural days would neceffarily be always of the fame length; and the apparent and mean time would be the fame. But experience has fhewn that this is not the cafe, and that the time which elapfes between the fun's being on the meridian of any place, and its return to it again, is confiderably longer fometimes than it is at others.

The annual motion of the earth is not perceived by us, who are upon it; but

it is the caufe of an apparent motion in
the fun, the fame way; namely, eaftward,
and of the fame quantity: Confequently,
when the earth, by its diurnal rotation
on its axis, has brought any place on its
furface oppofite to the point where the
fun was at the preceding noon, the in-
habitants of that place will not find the
fun there, but will have to follow it, ftill
farther eaftward, by a quantity which is
equal to the fun's apparent diurnal motion in its orbit, before the place they
inhabit will come oppofite to it. And, as
it has been obferved, this motion is not
only unequal in itfelf, but rendered, apparently, ftill more fo by the obliquity of its
direction, it is obvious the earth will have
to follow the fun fometimes a longer, and
fometimes a fhorter fpace before the fame
point on its furface will come oppofite to
it; and, of courfe, the lengths of natural
days will be fometimes longer, and fometimes fhorter alfo. But as all good clocks
and watches go uniformly, the mean day
of 24 hours, which is fhewn by them, muft,
neceffarily be always of the fame length:
it, therefore, follows, that when the fun's
apparent motion in its orbit is flow, and
the earth, in confequence, has a lefs fpace

to follow it before any given place on its
furface comes oppofite to the fun, the fun
will at fuch times be on the meridian of
that place before the end of the 24 mean
hours; and when the fun's apparent motion in its orbit is quickeft, and when, of
courfe, any given place on the earth's furface has a greater fpace to follow the fun
before it comes oppofite to it, the fun will
not be on the meridian of that place, till
fome time after the 24 hours are ended.

23. This inequality in the lengths of
natural days, is called the equation of time;
and the quantity of it, that is, the time by
which the fun appears to be on the meridian of any place before or after the end
of 24 mean hours, is inferted on the fecond page of every month in the Nautical
Almanac, for the noon of each day, at
Greenwich. It is marked *fubtractive*, when
the fun comes to the meridian fooner, and
additive when it comes to the meridian
later than the time of mean noon, or twelve
o'clock as it is ufually called: and the mean-
ing is, that the quantity of time exprefted
by this equation is to be fubtracted from
the apparent time, or that which is immediately derived from the fun, to obtain the
mean time, fhewn by clocks and watches,

in the former cafe, and added to it in the latter.

If the equation of time be taken from 24^h when it is, fubtractive, the remainder will be the mean time when the fun's center will be on the meridian of the place: when the equation is additive, the equation is itfelf, the mean time of apparent noon.

PROBLEM I.

24. To take the equation of time out of the Nautical Almanac, for any given place and time.

RULE.

If the time at Greenwich be not given, turn the longitude of the place into time; and add it to the time at the given place, if the longitude be weft, but fubtract it from that time, if the longitude be eaft, and it will give the time at Greenwich.

Take the equation of time from page II. of the Nautical Almanac, for the noon preceding the time when it is wanted, and alfo the difference between it and the equation for the day following; and fay, as 24^h is to this difference, fo is the time at Greenwich to a fourth number; which muft be added to, or fubtracted from, the equation for the preceding noon, according as the equation is increafing or decreafing.

Note. In every operation, where one time is to be taken from another, add 24 hours to the time you fubtract from, if the time which is to be taken from it be the greater; and the remainder muft be reckoned from the noon of the preceding day. When one time is added to another, if the fum exceed 24 hours, take 24 hours from it, and the remainder muft be reckoned from the noon of the following day.

EXAMPLE I.

What was the equation of time on the 27th of December, 1793, when it is 27'. paft 10 o'clock at Greenwich ?

Equation of time for noon, Dec. 27th, 1793, add. 1' 50", 5

24ʰ is to 29", 5 (daily diff.) as 10ʰ 27' is to   +   12 , 8

The equation fought +2   3 , 3

EXAMPLE II.

What was the equation of time on the 13th of November, 1793, at noon, at Calcutta, longitude in time 5ʰ 54' ?

Equation of time for noon at Greenwich . . . . 15' 21", 5

24ʰ is to 5ʰ 54' as 9" (daily diff.) to . . . . +   2 , 2

Equation of time for noon at Calcutta . . . . . . 15 23 , 7

EXAMPLE III.

What was the equation of time on the 14th of July, 1794, at 18ʰ 46′ at Mexico?

Time at Mexico, July 14th, ... 18ʰ 46′
Long. of Mexico in time 6 40

Time at Greenwich, July 15th, 1 26
Equation of time, July 15th, at noon, add...... 5′ 29″,6
24ʰ is to 5″,7 (daily diff.) as 1ʰ 26′ is to ..... + 0 , 3

The equation fought, add ..... 5 29 , 9

PROBLEM II.

25. To find the fun's longitude for noon at any given place from the Nautical Almanac.

RULE.

Take the fun's longitude from page II. of the Nautical Almanac for the given day, and its hourly motion from page III.

Enter Table III. with the fun's hourly motion at the top, and the longitude of the given place in the left-hand-fide column, and take out the correction which ftands under the former and oppofite the latter: this correction being added to the fun's longitude for noon at Greenwich, if the longitude of the place be weft, or fubtracted from it if the longitude be eaft, will give the fun's longitude for noon at the given place.

EXAMPLE I.

What was the fun's longitude at noon on the 15th of October, 1793, at Lima, in Peru?

Sun's longitude for noon at Greenwich   6ˢ 22° 34′,8
Hourly mot. 2′ 29″ and long. 77° W. give +   12,7

Sun's long. for noon at Lima   6   22   47, 5

EXAMPLE II.

What was the fun's longitude on the 27th of November, 1793, at Calcutta?

Sun's longitude for noon at Greenwich   8ˢ 5°46′,6
Hourly motion 2′ 32″ and long. 88ᵒ E. give —   14,9

Sun's longitude for noon at Calcutta   8   5   31 , 7

PROBLEM III.

26. To find the fun's declination for any given time and place from the Nautical Almanac.

RULE.

Take the fun's declination out of the Nautical Almanac for noon at Greenwich on the given day, if the given time be lefs than 12 hours, but for the day following if it be more.

Enter Table VI. of the Requifite Tables with the time from the neareft noon at the top, and the day of the month in one of the fide columns: under the former, and oppofite to the latter, ftands the correction of the fun's declination on account of the time.

If the fun's declination be wanted for noon at any other place than Greenwich, enter the Table with the longitude of the given place at the top, and the day of the month in one of the fide columns, againft which, and under the longitude, ftands the correction of the fun's declination on that account.

If the declination be wanted at any other place than Greenwich, and at any other time but noon, both corrections muft be applied; and they muft be added to, or fubtracted from, the declination for noon at Greenwich, according to the directions which ftand at the top of the column where the day of the month is found.

EXAMPLE I.

Find the fun's declination for 21^h $17'$ apparent time at Greenwich, May 4th, 1793.

May 4th, at 21^h $17'$ is 2^h $43'$ before noon on the 5th.

Sun's declination for noon, May 5th, $16°$ $27'$ $34''$ N.

2^h $43'$ before noon gives . . . . — 1 59

___

Sun's declination, May 4th, at 21^h $17' = 16$ 25 35 N.

EXAMPLE II.

What was the fun's declination on the 19th of Auguſt, 1793, at noon, in longitude 97° weſt ?

Sun's declination Auguſt 19th, at noon ... 12° 34′ 35″ N.
97° Weſt longitude in Table VI. give ... — 5 30

Sun's decl. Aug. 19, at noon, in long. 97° W. 12 29 5 N.

EXAMPLE III.

What was the fun's declination on the 14th of October, 1793, at 7ʰ 34′ in longitude 83° eaſt ?

Sun's declination October 14, at noon ..... 8° 25′ 26″ S.
7ʰ 34′ afternoon in Table VI. give .... + 6 54
83° Eaſt longitude in Table VI. give ..... — 5 3

Sun's declination at 7ʰ 34′ in long. 83° eaſt 8 27 17 S

PROBLEM IV.

27. To find the mean time at any place, the latitude and longitude of that place; or the latitude, and time at Greenwich, being given.

RULE.

Let ſeveral altitudes of the fun's lower limb be obſerved when it is, at leaſt, three or four points of the compaſs from the meridian. Add theſe altitudes together, and divide the ſum by the number which were taken, and the quotient will be the mean obſerved altitude.

From the obſerved altitude of the fun's lower limb, ſubtract the dip of the horizon,

(Table II. of the Requifite Tables) and the
refraction (Table I.) and add the fun's
femidiameter, (page III. of the Nautical
Almanac), and the parallax in altitude
(Tab. III.) to the remainder; the fum will
be the true altitude of the fun's center.

Take the fun's declination from page II.
of the Nautical Almanac by Problem III.

If the fun's declination and the co-
latitude of the fhip be one north and the
other fouth, take their difference; but if they
be both north, or both South, take their
fum for the fun's meridional altitude. If that
fum be greater than 90° take it from 180°.

From the natural fine of the fun's meridional altitude (Table XVII.) take the
natural fine of the fun's true altitude at
the time of obfervation; find the logarithm
of the remainder, to which add the logarithmic fecant of the fhip's latitude, and
the logarithmic fecant of the fun's declination; their fum, rejecting 20 from the
index, muft be fought for in Table XVI.
under log-rifing, and the time correfponding to it will be the apparent time from
the nearest noon: confequently, if the obfervation be made in the morning, the time
thus found muft be taken from 24 hours,
and the remainder will be the apparent
time from the noon of the preceding day.

Take the equation of time from page II. of the Nautical Almanac by Problem I. and add it to, or subtract from the apparent time as its title directs, and it will give the mean time.

EXAMPLE I.

January the 3d, 1793, about 20ʰ 45′ latitude 16° 23′ north, longitude 115° east, the following altitudes of the sun's lower limb were observed: what was the true mean time; the height of the observer's eye being 20 feet?

ALT. ⊙'s L.L.	
27° 45′	Sun's declin. Jan. 4th, noon 22° 39′ 1″ S.
27 51	3ʰ 15′ before noon give ... + 50
28 1	115° East longitude give + 1 58
28 7	Sun's d. at 20ʰ ¾ in l. 115° E. 22 41 49 S.
28 10	Sun's semidiameter 16 19
	Sun's parallax in altitude ... 8
139 54	16 27
———	Dip of the horizon 4′ 16″ } 6 3
27 58,48	Refraction ...... 1 47 }
+ 10,24	Correction sun's altitude 10 24
28 9,12	Sun's true altitude.

Ship's latitude   16 23′ N. Secant 10.01800
Co-latitude   73 37 N.
Sun's declin.   22 42 S. Secant 10.03502
Sun's merid. alt.   50 55 Nat. sine 77623
Sun's true obf. alt. 28   9 Nat. sine 47178
        Diff. nat. sines 30445 Log. 4.48351
Time before noon the 4th, 3° 16′ 01″ Log-rising 4.53653
Appa. time, January 3d, 20 43 59
Equation of time ...... + 5 32
        Mean time 20 49 31

EXAMPLE II.

February the 3d, 1793, about half-paſt two o'clock, latitude 15' 50' S. longitude 162° E. the following altitudes of the ſun's lower limb were obſerved: what was the true mean time; the height of the obſerver's eye being 21 feet?

ALT. ☉'S L. L.					

ALT. ☉'S L. L.	Sun's declin. Feb. 3d, at noon 16° 17' 5" S.
54° 30' 30"	2ʰ 30' afternoon, give .... — 1 46
54 25 00	162°. Eaſt longitude, give + 7 36
54 14 00	Sun's dec. at 2ʰ 30' in l. 162° E. 16 22 55 S.
54 03 00	Sun's ſemidiameter 16 16
53 57 00	Sun's parallax in altitude .. 5
53 49 00	16 21
324 58 30	Dip. of the horizon 4' 22" }
	Refraction ...... 0 41 } 5 3
54 09 45	
+ 11 18	Correction ſun's altitude 11 18
54 21 3	Sun's true altitude.

90°.

Latitude .... 15 50 S. Log-ſecant 10. 01680

Co-latitude ..... 74 10 S.
Sun's declination 16 23 S. Log-ſecant 10. 01800

90 33
180

Sun's merid. alt. 89 27 Nat. ſine 99995
True obſerved alt. 54 21 Nat. ſine 81259

18736 Log. 4. 27268

Apparent time Feb. 3d, 2ʰ 28' 37" Log-riſing 4. 30748
Equation of time ... + 14 18

Mean time 2 42 55

28. Having now fhewn how to find the mean time at the place where the obferver is fituated, I fhall proceed to fhew how it may be found at Greenwich, or the firft meridian.

If a watch, or time-keeper, be regulated to keep mean time, exactly, and be fet to the mean time at the firft meridian; it is manifeft that fuch watch will continue to fhew the mean time at that meridian, as long as it continues to go at the fame rate, whatever place it may be carried to: and, confequently, if a watch, fo regulated, be kept on board a fhip, it will always fhew the mean time at the firft meridian. Hence, if the mean time be found at the fhip, by the preceding problem, the difference between it and the time fhewn by the watch, when the fun's altitude was obferved, being turned into degrees and minutes at the rate of 15° to an hour, will be the longitude of the place where the fun's altitude was obferved, by Art. 18.

29. It is not, however, abfolutely neceffary that the watch fhould either be fet precifely to mean time at the firft meridian, or be regulated to keep exactly mean time; both of which might, perhaps, be difficult, or, at leaft, tedious to effect. The only

thing which is abfolutely requifite; in a watch, to render it equal to the tafk of finding the longitude, is, that it will go uniformly at fome rate; becaufe the rate which it does go at, as well as its deviation from mean time at the firft meridian, may be readily found, as follows, and allowed for.

30. Strictly fpeaking, the *rate of a watch* is the number of hours, minutes, and feconds, which its hands appear to have moved over on the dial-plate in the fpace of a mean folar day; but it is cuftomary to call the difference between this time and 24 hours the *rate of the watch*. If the time fhewn by the watch in a mean folar day be more than 24 hours, 24 hours are taken from it, and the remainder is called the *rate of the watch*; which is marked +, fignifying that the watch gains. If the time fhewn by the watch in a mean folar day be lefs than 24 hours, that time is fubtracted from 24 hours, and the remainder is called the *rate of the watch*; and is marked —, which fignifies that the watch lofes. And this is the fenfe which the term is to be underftood in throughout this tract.

PROBLEM V.

31. To find the rate which a watch goes at; that is, how much it gains or lofes on mean time in a day, or 24 hours; and how much it is too faft, or too flow, for mean time, at any place.

RULE.

The rate which a watch goes at may be found very readily, and with tolerable exactnefs, without the ufe of any inftrument but Hadley's quadrant, or any obfervations but that of the fun's altitude, which every feaman is, neceffarily accuftomed to make.

Sometime before the watch is wanted, a month may be fufficient in moft cafes, obferve feveral altitudes of the fun's lower limb, when it is, at leaft, four points of the compafs from the meridian, and note the times when they were obferved by the watch.

Add thefe altitudes together, divide the fum by the number of obfervations, and the quotient will be the mean altitude of the fun's limb. Take, in like manner, the mean of the times by the watch.

Correct the mean obferved altitude for the dip of the horizon, refraction, the fun's femidiameter, and its parallax in altitude; and, with the altitude fo corrected, the latitude of the place, and the fun's declination find the mean time by Problem IV.

Take the difference between the mean time, and the mean of the times fhewn by the watch, when the obfervations were made, and it will be what the watch was too faft or too flow for mean time.

Immediately before the fhip fails, repeat thefe obfervations; and find again how much the watch is too faft or too flow for mean time. Take the difference between the mean time of the firft obfervation, and the mean time of the fecond; alfo the difference between what the watch was too faft, or too flow, at thefe two times; and the latter difference will be the lofs or gain of the watch, in the interval between the obfervations. Then, as the interval between the obfervations is to one day, fo is the lofs or gain of the watch, between the obfervations, to its lofs or gain in one day.

EXAMPLE.

March 27th, 1793, the following obfervations were made on board a fhip, lying in the Downs, latitude 51° 13′ N. and long. 1° 27′ E. for determining the rate which a

watch went at, the height of the obferver's
eye being 22 feet.

Times by the watch.	Alt. of the Sun's L.L.		
		Time at the fhip ... 19ᵇ 2′	
		Longit. in time eaft 6	
		Time at Greenwich 18 56	
19ᵇ 0′ 52″	11° 13′ 30″		
1 26	19	The fun's femi-diameter 16′ 4″	
1 50	25 30	The fun's parallax in alt. 9	
2 21	31 15		
2 44	37	16 13	
9 13	126 15	Dip of the hor. 4′ 28″ ⎫ 9 6	
		Refraction 4 38 ⎭	
19 1 50,6	11 25 15		
	+ 7 7	Correction of the ☉'s. alt. 7 7	
	11 32 22	Sun's true altitude.	

3°17′ 5″ N. fun's declin. noon, on the 28th.
— 4 56 Correction for 5ᵇ 4′ before noon.

Sun's declin. 3 12 9 N. Secant 10.00068
Co-latitude 38 47 N. Co-fecant 10.20316

Merid. Altit. 41 59 9 N. S. 66894 ⎱ 46890 Log.4.67108
True Altit. 11 32 22 N. S. 20004 ⎰

5ᵇ 2′ 2″,0 Log-rifing 4.87492
24

Apparent time 18 57 58,0
Equation of time .. + 5 04,0

Mean time 19 3 02,0
Time by the watch 19 1 50,6

Watch too flow.... 1 11,4

32. Again, May 3d, 1793, the fhip being ready to fail, the following obferva·

tions were made: the error of the quadrant being 1′ 53″ to be added.

Times by the watch.	Alt. of the Sun's L. L.	
18ʰ 21′ 37″	16° 16′ 45″	Time at the ship . . . 18ʰ23′
22 13	22	Longitude in time . . 6
22 44	26 30	Time at Greenwich 18 17
23 17	31 45	Sun's femidiameter . . . . 15′ 54″
23 50	37 15	Parallax in altitude 8
		Error of the quadrant 1 53
		17 55
13 41	134 15	Dip of the hor. 4′ 28″ ⎫ 7 40
		Refraction 3 12 ⎭
18 22 44, 2	16 26 51	
	+ 10 15	Correction fun's altitude 10 15
	16 37 6	Sun's true altitude.

16° 10′ 31″ N. fun's de. noon at Greenwich.
— 4 07 Correction for 5ʰ 43′ bef. noon.

Sun's declin. 16 6 24 N. Secant 10. 01739
Co-latitude 38 47 00 N. Co-fecant 10. 20316

Meridian altit. 54 53 24 N.S. 81805 ⎫ 53205 Log. 4. 72595
True altitude 16 37 6 N.S. 28600 ⎭

5ʰ 33′ 22″,6 Log-rifing 4. 94650
24

Apparent time 18 26 37, 4
Equation of time . . . — 3 29, 6

Mean time 18 23 7, 8
Time by watch 18 22 44, 2

Watch too flow . . . . 0 23, 6 at 18ʰ 23′ 8″ m. T.
Watch too flow March 27th 1 11, 4 at 19 3 02 m. T.

Watch has gained . . . . , 0 47, 8 in 36ᵈ 23 2Q 6, or
1″,293 in a day.

33. To obtain the rate which the watch goes at with greater certainty, obfervations may be made on more days than one, both at the beginning and end of the interval: or more than one fet of obfervations may be taken on the fame day, if opportunities offer, and the refult of each fet of obfervations, made at the beginning of the interval, muft be compared with the refult of each of thofe taken at the end of it; by which means feveral daily rates will be obtained, which may differ a little one from another. In this cafe, the mean of them all muft be taken for the true rate which the watch goes at.

34. Obfervations made only at the beginning and end of the interval, give the rate of the watch, on a fuppofition that it does not alter its rate of going in the time which elapfes between them; for if it does, thofe obfervations will not detect it. It will therefore be prudent to take obfervations as often as circumftances will permit, and find the rate which the watch has gone at between every two obfervations.

35. I have given this method of finding the rate which a watch goes at, becaufe it may be put in practice by every feaman, without introducing the ufe of any

inftrument, or obfervation, which he is not already neceffarily acquainted with ; and be-caufe it admits of being executed, if care and fkill be exerted, with tolerable exactnefs : fufficient, at leaft, for the ufual length of a Weft-India voyage, out, or home. It how-ever, requires a very good inftrument, and care and fkill in the ufe of it : moreover, the utmoft exactnefs muft be obferved in the calculations ; and, when the utmoft fkill in both is exerted, it is not fufficient for long voyages.

36. The moft exact way of finding the rate which a watch goes at, and that which requires the leaft calculation, is the method practifed in fixed obfervatories, where they have tranfit inftruments adjufted to, and moving in the meridian of the place. This method has been lately adopted by fome gentlemen *, in the Eaft-India Company's fervice, who have carried out portable tran-fit inftruments, for the purpofe of examin-ing their watches in India, before they fet out on their voyage home.

* It was Jofeph Huddart, Efq. F. R. S. an Elder Brother of the Trinity-Houfe, and lately Commander of the Royal Admiral Eaft-Indiaman, a very ingenious aftronomer and me-chanic, who, I believe, introduced this bufinefs into the fer-vice.

DESCRIPTION AND USE OF THE TRANSIT INSTRUMENT.

37. A tranfit inftrument is a telefcope, of any convenient length, fixed at right angles to an horizontal axis; on which it turns in the plane of the meridian, or any other vertical circle.

Let F E *(fig. 2,)* reprefent an horizontal axis, the two ends of which, F and E, are perfect cylinders, and reft on the frame, or fupport, A B C D, in the angular notches A and D. The frame A B C D may be of caft-iron; but the notches, or Y s, A and D, muft be of bell-metal, and flide in grooves, which are fcrewed firmly to the two upper ends of the iron frame. That at A has a vertical motion, governed by the fcrew *h*, by which the end F, of the axis E F, is raifed, or lowered, to bring the axis horizontal. The Y at D has an horizontal motion, governed by the fcrew *g*, to bring the telefcope to any mark the obferver may wifh to place it to; or to reftore it to that mark if it be found, at any time, to deviate from it. Thefe Y s, or notches, with the apparatus which governs them, are reprefented more diftinctly in *fig.* 3 and 4, where S reprefents that which has the vertical motion at A; and T that which has the horizontal motion at D. F and E are fections of the axis, E F, refting in them.

38. The axis E F (exclufive of the two cylindric pivots, E and F,) is formed of two hollow conic fruftums of brafs, or tin; the two greater bafes meeting at W, where they are croffed, at right angles, by two other hollow conic fruftums, W G, W H, through which the telefcope M N paffes; the apertures at G and H being fomewhat wider than is neceffary to receive it. To fix the telefcope in its place, there are four fmall fcrews, two at the aperture G and two at H. Thofe at G (toward the eye-end of the telefcope) act one againft the other, in an horizontal direction; one of them appears at *b*, and the other is diametrically oppofite to it, and, confequently, hid by the telefcope. The two fcrews which are at H, (one of which appears at *c*) act one againft the other in a vertical pofition. Thefe fcrews have fine conical points; fo that when they are forced againft the tube of the telefcope, they will form fmall conical holes in it for themfelves to reft in.

39. Five very fine filver wires are placed in the common focus of the two glaffes of the telefcope, exactly parallel to, and equidiftant from each other: the middle one bifecting as nearly as the inftrument-maker can poffibly make it, the field of the telefcope; and a fixth wire is placed at right angles to thefe, bifecting alfo the field of the telefcope. This fyftem of wires is reprefented in *fig.* 5, and requires no farther explanation.

40. O P Q R is a fpirit level, which being hung on the two cylindric ends of the axis F and E, by means of the angular hooks, O and R, the end F, of the axis, is raifed or lowered by the fcrew *h*, until the level fhews that the axis is horizontal.

41. L Z K I is a femicircle, divided into two quadrants of 90 degrees each, beginning at L and I, and meeting in Z, at 90°. E K is an index, twifted on to the end E of the axis, having a Vernier's divifion at K, which fubdivides the arch to every 3′; and which is fet to 90°, at Z, when the telefcope points to the zenith. By means of this arch and index the telefcope may be fet with precifion, to the meridional altitude of any object, the paffage of which,

over the meridan, is required to be obferved; whether it paffes to the northward or
fouthward of the zenith.

42. The cylindric pivot F is perforat-
ed, and fo alfo is that fide of the tube of
the telefcope, which is oppofite to it; and
within the tube, and directly oppofite this
perforation, a plane reflector is fixed, at an
angle of 45° with the axis of the telefcope,
having a hole through it, large enough to
admit all the rays paffing from the object
glafs of the telefcope.

43. When ftars are obferved in the
night, a fmall candle, in a ftick which has
a contrivance for fetting the candle higher
or lower, is placed oppofite the hole in
the pivot, the light of which falling on
the reflector in the tube of the telefcope,
is thrown by it on the wires in the focus
of the telefcope; by which means they are
rendered vifible to the obferver. By placing the candle nearer to, or farther from
the end of the pivot F, the quantity of
light thrown on the wires may be regulated to the wifh of the obferver.

TO LEVEL THE AXIS.

44. Hang the level on the axis, the
hook O on the pivot F, and the hook R

on the pivot E, and raife or lower the end
F, by the fcrew *h*, until the bubble refts
exactly in the middle of the glafs tube.
This being done, invert the level, hanging
the hook O on the pivot E, and the hook
R on the pivot F. If the bubble ftill reft
in the middle of the tube, the axis of the
inftrument is parallel to the horizon. But
if the bubble do not now reft exactly in
the middle of the glafs tube of the level,
turn the fmall fcrew *q*, which is at the end
Q of the level, until you have moved the
bubble half-way toward the middle of the
tube, and carry it the other half by turning the fcrew *h*. Invert the level again,
hanging the hook O on the pivot F, and
the hook R on the pivot E; and if you
were exact in eftimating half the error of
the level, the bubble will now reft exactly
in the middle of the tube, and the axis will
be level. But if it be not, correct half the
error by the fcrew *q*, and the other half by
the fcrew *h*, and invert the level until
the bubble will reft exactly in the middle of the tube, when the level is hung on
both ways.

TO FIX THE TELESCOPE AT RIGHT ANGLES TO THE AXIS.

45. Let the fcrews at G and H be withdrawn till their points be barely even with the internal furfaces of the two conical arms W G, W H, lay the axis in the Y s, and place it truly horizontal by means of the level. Turn the axis in the Y s until the two arms W G, W H, are alfo in an horizontal pofition, pafs the telefcope through thefe arms until it be exactly in equilibrio on the axis, and then let it reft freely in the arms with the five parallel wires, as nearly perpendicular to the horizon as you can place them by your eye. Turn the frame which carries the inftrument round on the fcrew-pin U, until the middle wire, of the five parallel ones, be on fome diftant, well-defined object, and turn the telefcope a little on the axis E F, noting at the fame time, whether the middle wire remains conftantly on the mark throughout the whole extent of the field of the telefcope. If it does not, the telefcope muft be turned in the arms W G, W H, until the middle vertical wire will remain exactly on the mark, when the inftrument is turned on the axis E F, through-

but the whole extent of the field. Turn now the ſcrew *c*, till its point preſſes the tube of the teleſcope, as it lies in the arms W G, W H; taking care that the teleſcope ſtill remains in the ſame poſition, and that the middle wire will remain exactly on the mark throughout the whole extent of the field of the teleſcope, at the ſame time that the axis F E is ſhewn by the level to be parallel to the horizon. Being well aſſured of theſe two points, turn the ſcrew *c*, till its point, by preſſing againſt the tube of the teleſcope, forms a ſmall conical impreſſion in it; and this being done, turn the ſcrew which is oppoſite to *c*, till its point alſo make a ſmall impreſſion in the tube of the teleſcope; after which the ſcrew *c* may be eaſed, ſcrewing up the oppoſite one at the ſame time, till the teleſcope is brought into the center of the aperture at H. And it is plain, that if the wires have been kept, during this operation, perpendicular to the horizon, they muſt remain ſo, as long as the ſcrews are kept tight, becauſe the teleſcope can have no power to turn round. Raiſe the end M of the teleſcope, in the aperture G, by putting ſlips of paper under it, until it be in the center of the aperture: then turn

the screw *b*, and its opposite one, until
their points press against, and form small
conical holes in the tube of the telescope.
By turning the screw *g*, bring the middle
wire, of the five vertical ones, very exactly on to the mark which was used in
setting it perpendicular to the horizon, or on
to any other which is small, and perfectly
distinct: then turn the axis end for end;
that is, turn the pivot E into the notch A,
and the pivot F into the notch D; and if
the middle vertical wire be still exactly on
the mark, the line of collimation of the
telescope is perpendicular to the axis F E.
If the middle wire be not on the mark
when the axis has been inverted, bring the
wire half way toward the mark by easing
one of the screws at *b*, and screwing up the
other; and then bring the wire exactly on
to the mark by the screw *g*. Invert the
axis EF again; that is, bring the end F
into the notch A, and the end E into the
notch D; and, if you were exact in esti-
mating half the distance which the wire
deviated from the mark, it must now be
exactly on it. If it be not, bring the wire
half way toward the mark by the screws
at *b*, and the other half by the screw *g*;
and try, again, by inverting the axis, whe-

ther you have now done it right; if you
have not, the operation muſt be repeated
until the middle wire be exactly on the
mark, which ever end of the axis FE is
in the notch A.

TO ADJUST THE TELESCOPE.

46. That is, to place the eye-glaſs and
object glaſs at ſuch a diſtance from each
other, that their reſpective *foci* may co-incide:
after which the wires are to be brought
into their common focus. To effect this,
ſome teleſcopes have the eye-glaſs and cell
which carries the wires moveable, while
the object-glaſs is fixed: others have the
wires fixed and the two glaſſes moveable.
In the former caſe, by puſhing in, or draw-
ing out the eye-piece, adjuſt the teleſcope
ſo that the ſun or a planet appears perfectly
diſtinct through it; then move the wires
nearer to, or farther from the eye-glaſs, as
may be required, until they alſo appear
perfectly diſtinct, and the teleſcope will be
adjuſted ready for uſe. In the latter con-
ſtruction, puſh in, or draw out the eye-
piece till the wires appear perfectly diſtinct,
then alter the object-glaſs until the ſun, or
a planet, appear perfectly diſtinct alſo, and

the telefcope will be adjufted ready for
ufe.

47. As it is of importance to have the
telefcope adjufted very exactly in this
refpect, the following method of trying
whether it be fo or not, may be practifed.

The telefcope being adjufted to diftinct
vifion for diftant objects, when a fixed ftar
is on the meridian, bring the horizontal
wire to bifect it very exactly, and the ftar
will run along the wire through the whole
extent of the field of the telefcope. While
the ftar is thus running along the wire,
move your eye a little upward or down-
ward; and, if the wires be not exactly in
the common focus of the two glaffes, the
ftar will appear to quit the wire when the
eye is moved. If this be the cafe, the wires,
or glaffes, muft be altered until the ftar
will not quit the wire when the eye is
moved; the objects appearing perfectly diftinct at the fame time.

TO OBSERVE THE PASSAGE OF A STAR OVER THE MERIDIAN BY A TRANSIT INSTRUMENT.

48. From the latitude of the place and
the ftar's declination find its meridional al-

titude *; and a little before the time when
it will be on the meridian †, set the telescope to this altitude, by means of the
semi-circle L Z I, and index E K. Apply
your eye to the telescope, and wait till you
see the star come into it, which will be
(apparently) on the west side, because the
telescope inverts objects. Let your assistant
attend to the watch, and at the instant
when the star is on the first wire, bid
him *mark* the second, and parts of a second,
when it happened; which he must set
down on a paper, ready ruled for that purpose, and then attend again to the watch.
When the star is on the second wire, bid
him *mark* the second, &c. when it happened,
as before; which being written down, with
the minute, proceed to observe the instants
when the star is on every one of the other
wires; after which the proper hour may
be annexed.

An assistant is not necessary when the
time is taken by a clock, the beats of which
can be distinctly heard by the observer:
indeed, the observation may be made more

* If the latitude and declination be both north, or both
south, add the co-latitude to the star's declination : but if one
be north and the other south, take their difference.

† This may be found by subtracting the sun's right af-
cension for the day, (Nautical Almanac, p. II.) from the star's
right ascension, (Requisite Tables, Tab. VII.)

exactly without one. When a clock is ufed, the obferver, catches the fecond from the clock by his eye, which he then applies to the telefcope, telling the feconds on by his ear until he fees the ftar on the wire. He then writes down the fecond and parts of a fecond when it happened, and prefixes the minute; after which, he catches the fecond from the clock again by his eye, and tells on by his ear until the ftar arrives at the fecond wire, and fo on.

TO BRING A TRANSIT INSTRUMENT INTO THE MERIDIAN.

49. Take altitudes of the fun, noting the times by the watch when they were taken: and, from thence, the latitude of the place, and the fun's declination, find the *apparent* time by Art. 27. The diffe-rence between this time, and the mean of the times fhewn by the watch when the ob-fervations were made, will be what the watch is too faft, or too flow, for *apparent* time.

If the watch be too faft, add what it is too faft to 12 hours: but if it be too flow, fubtract what it is too flow from 12 hours, and you will have the time by the watch, when the fun will be on the meridian, as near as the going of the watch can be depended on. Take the time which the fun's femidiameter is paffing the meridian from page III. of the Nautical Almanac, and add it to, and fubtract it from the time by the watch, when the fun will be on the meridian, and you will have the times when the fun's eaftern and weftern limbs will be on the meridian.

A few minutes before the time when the weftern limb will be on the meridian, bid your affiftant tell the feconds, as they pafs, by the watch; but inftead of calling the 60th fecond, let him name the minute the watch is then at. While he is doing this you muft bring the fun into the tele-fcope by elevating it to the proper altitude; and turning the whole inftrument round on the fcrew-pin U. Having, by this means brought the middle wire apparently to the eaftward of what appears to be the eaftern limb of the fun, (becaufe the fun will ap-pear to move that way in the telefcope) tighten the fcrew U by turning the nut; and when the fun's limb arrives at the mid-dle wire, keep it on it by turning the fcrew g, at the rate the fun moves, till your affiftant calls the fecond by the watch at which you had computed the weftern limb

G

of the fun would be on the meridian, and the inftrument will be nearly in the meridian. Let your affiftant tell on till the watch arrives at the fecond when, according to your calculation, the eaftern limb of the fun fhould be on the meridian; and, if it is not exactly on it, you will have another opportunity of rectifying the inftrument by turning the fcrew g.

50. On this, and feveral other occafions, if you have not a darkening-glafs fitted to the eye-end of the telefcope, it will be convenient to have a pair of fpectacles with plane dark-glaffes, of different degrees of darknefs, inftead of lens.

51. Having thus brought the inftrument into, or very near the meridian, its real fituation, with refpect to the meridian, may be verified feveral ways; of which I fhall point out two. If the latitude of the place be confiderable; that is, 30 degrees, or upward, there are a variety of .ftars in both hemifpheres, fufficiently bright, which never fet; and, confequently, may be obferved at the inftrument both above and below the pole. Let the tranfits of fuch a ftar over the meridian be obferved above the pole, below the pole, and above the pole again : and it is manifeft that if the

time of the firſt tranſit above the pole, be
ſubtracted from the time of the ſecond tranſit above the pole, (adding 24 hours, if
neceſſary) the remainder will be the time
by the watch, in which the earth (or the
ſtar apparently) makes one diurnal revolution. It is alſo evident, that if the two
intervals between the time of the tranſit
below the pole, and the two tranſits above
the pole be equal, the inſtrument muſt be
exactly in the meridian. If the interval
between the firſt tranſit above the pole,
and the tranſit below the pole, be greater
than the interval between the tranſit below
the pole, and the ſecond tranſit above it,
the object end of the teleſcope, when directed toward the elevated pole, lies to the
eaſt of the true meridian; but if the latter
interval be greateſt, the object end of the
teleſcope, when directed toward the elevated pole, lies weſt of the true meridian.
To correct the error, and bring the inſtrument into the meridian, add 24 hours to
the time of the latter tranſit above the
pole, ſubtract the time of the former from
it, and take half the remainder. Take the
difference between this and the interval
between the tranſits above and below the
pole, and take half this difference. Then

as the time by the watch of an entire revolution is to 24 hours, so is this half difference to the half difference in siderial time.
Add together the logarithm of this half-
difference, the logarithmic tangent of the
star's polar distance; and the logarithmic
secant of the latitude of the place, the
sum, rejecting 20 from the index, will be
the logarithm of the number of seconds *in
time*, which expresses the angle made by the
instrument and the meridian.

Consider what part this angle makes of
the interval, between the wires which are
in the focus of the telescope *: and turn
the instrument, on its axis, till the telescope points to the horizon. Look out for
some tolerably distinct object which is cut
by one of the wires; and, by turning the
screw *g*, remove the wire to the east or west
of this object, (as may be required) such a
part of the space between, that wire and
the next to it, as the angular error of the
instrument makes of that interval. You
must then proceed to examine the position

* The interval between the wires in the focus of the telescope may be found by observing the transits of one, or more
of the stars δ Ceti, δ Orionis, υ Leonis, ϖ or γ Virginis, ϖ An-
tinoi, λ Aquarii, κ or λ Piscium, or any other star which is
very near the equator.

of the inftrument again, either by the fame, or fome other circum-polar ftar, and to cor- rect it, if it requires correction, until you get it exactly into the plane of the meri- dian : and when you have, a mark muft be fet up in the meridian, at as great a dif- tance from the inftrument as may be con- venient, or as it can be feen diftinctly ; and the telefcope muft be carefully adjufted to this mark before every obfervation.

E X A M P L E.

52. Admit that Capella paffed the tranfit inftrument, at Greenwich, Septem- ber 6th, 1794, above the pole at 17^h $59'$ $11'',5$, below the pole at 5^h $57'$ $5'',6$, and above the pole, again, at 17^h $55'$ $21'',0$: how much did the inftrument deviate from the meridian, and which way ?

The difference between the times of the firft tranfit above the pole and the fe- cond is 23^h $56'$ $9'',5$; half of which is 11^h $58'$ $4'',75$. The difference between the times of the firft tranfit above the pole and the tranfit below it, is 11^h $57'$ $54'',1$; which is lefs than the time of half a revolution by $10'',65$; and half of this is $5'',325$. Hence, the object-end of the telefcope, when di- rected toward the elevated pole was weft

of the meridian. As 23ʰ 56′ 9″,5 (the time
of a revolution of the fixed ſtars by the
watch) is to 24 hours, ſo is 5″,325 to 5″,34.
And as the tangent 45° 46′ 25″, the ſtar's
declination, is to the ſecant of 51° 28′ 40″,
the latitude of the place, ſo is 5″,34 to
8″345, the angle under which the inſtrument cuts the plane of the meridian, in
time.

53. To adjuſt the inſtrument to the
plane of the meridian, let us ſuppoſe that
a ſtar at the equator had been obſerved,
and found to take up 36″¼ in paſſing from
wire to wire, 8″,345 is about two-ninth
parts of this quantity, or ſomewhat more
than one-fourth. Bring the teleſcope to
the horizon, and look out for ſome diſtinct
object which is cut by one of the wires;
then, by turning the ſcrew *g*, remove the
teleſcope until the object which was on the
wire, appears to be about two-ninths, or
nearly one-fourth of the diſtance between
that wire and the next to it toward the
right hand: if the inſtrument had been
eaſt of the meridian, the ſcrew *g* muſt
have been turned till the object appeared
to be removed, the like quantity toward
the left-hand. In conſidering the propriety
of theſe inſtructions, the reader muſt re-

member that the telefcope inverts objects.

54. This is the beft practical method
that I know, for placing a tranfit inftrument in the meridian, when it can be ufed:
but if the place of obfervation be very near
the equator; or, if any local impediment
prevent you from obferving circum-polar
ftars above and below the pole, the following one, given by the Rev. Dr. Mafkelyne,
Aftronomer Royal, in the Philofophical
Tranfactions for 1775, Vol. LXV. may be
ufed. But before this can be done it will
be neceffary to fhew how to find the deviation of a fixed ftar in right afcenfion in
time.

55. Take the ftar's right afcenfion in
time, and its declination, from Table IV. for
the time when it is wanted.

Turn the right afcenfion in time into
motion by allowing one fign for every two
hours, 15 degrees for one hour, one degree
for every four minutes, and one minute for
every four feconds of time.

Enter Table V.* with the ftar's right
afcenfion, and take out the degrees and

* This Table is Table VI. in Dr. Mafkelyne's collection,
adapted to the purpofe under confideration.

minutes correfponding to it, and alfo the logarithm.

Add the degrees, &c. taken out of Table V. to, or fubtract them from the ftar's right afcenfion as the figns + or — direct; and take the difference between the right afcenfion, fo corrected, and the longitude of the moon's afcending node, (Nautical Almanac, p. III.) fo that the remainder may be lefs than fix figns, or 180 degrees.

To the logarithmic co-fine of this remainder add the logarithm taken out of Table V. and the logarithmic tangent of the ftar's declination : their fum, rejecting all the tens in the index, will be the logarithm of the ftar's deviation in decimal parts of a fecond. To be added to the right afcenfion in time of a northern ftar, when the difference between the ftar's corrected right afcenfion and the longitude of the moon's node is greater than 3 figns, or 90°; but it muft be fubtracted from it when that difference is lefs than 3 figns. If the ftar have fouth declination, the deviation muft be fubtracted from the right afcenfion, when the difference between the ftar's corrected right afcenfion and the longitude of the moon's node is greater than 3 figns, and added to it when that difference is less than 3 signs.

EXAMPLE I.

What is the deviation of Fomalhaut, September 28th, 1796?

Star's Æ at the time   22ʰ 46′ 24″ = 11ˢ 11° 36′
Correction from Table V. . . . . + 4 30

Star's corrected right ascension 11 16 06
Long. ☽'s node (Naut. Alm. p. III.) 15 6 13

Difference. . . . . 3 20 07 cos. 9. 536
Logarithm from Table V. 9. 795
Star's declination south . . . . tang. 9. 774

Fomalhaut's deviation . , ... subt. 0″,13 log. 9. 105

EXAMPLE II.

What is the deviation of β Pegasi, September 28th, 1796?

Star's Æ at the time 22ʰ 53′ 56″ = 11ˢ 13° 29″
Correction from Table V. . . . . + 4 4

Star's corrected right ascension 11 17 33
Long. ☽'s node (Naut. Alm. p. III.) 15 6 13

Difference . . . . 3 18 40 cos. 9. 505
Logarithm from Table V. 9. 797
Star's declination north . . . . tang. 9. 707

Deviation of α Pegasi . . . . 0″,1 log. 9. 009

56. Dr. Maskelyne's method is general, but to avoid some parts of the calculation

H

which feamen, in general, are not accuftomed to, I fhall give it with certain limi-
tations.

Chufe two ftars which have nearly the
fame right afcenfion : let their declinations
alfo be nearly equal, neither of them lefs
than 25 or 30 degrees; but let one be north,
and the other fouth, fo that their difference in declination may not be lefs than
50 or 60 degrees. Moreover, let them not
pafs the meridian before 9 o'clock, nor after
three. Obferve the tranfit of each of them
twice by the inftrument; from which you
will obtain the time fhewn by the watch
during a revolution of the fixed ftars, and
alfo between the tranfits of the two ftars
over the inftrument. Reduce the time
elapfed between the tranfits of the two
ftars, according to the watch, to parts of
the equator, and it will be the obferved
difference of their right afcenfions. Take
the right afcenfions and declinations of the
two ftars from Table IV. for the time of
the obfervation, correct their right afcenfions for deviation*, by Art. 55, and take

---

* The mean places of the ftars are exhibited in the tables ;
and to obtain their apparent places, which are thofe obferved,
the mean places muft be corrected for aberration, and the equation of the equinoctial points, as well as for deviation : but as

the difference of the right afcenfions, fo
correfted; and if this difference be equal
to the obferved difference of their right
afcenfions, the inftrument is truly in the
meridian. If the obferved difference of
their right afcenfions be greater than the
true, and that ftar pafs firft which is neareft
the elevated pole; or, if the obferved difference of right afcenfion be lefs than the
true, and the ftar neareft the elevated pole
pafs laft, that end of the telefcope which
is directed toward the elevated pole is eaft
of the meridian. If the obferved difference
of right afcenfion be lefs than the true,
and the ftar neareft the elevated pole pafs
firft; or if the obferved difference of right
afcenfion be greater than the true, and the
ftar neareft the elevated pole pafs laft,
that end of the telefcope which is directed
toward the elevated pole is weft of the meridian.

57. Having thus found which way the
inftrument deviates from the meridian, the

it is only the difference between the apparent right afcenfions of
the two ftars which is wanted here, and not the right afcenfions themfelves; and, as the equation of the equinoctial
points affects the right afcenfions of all ftars alike, and the
ftars in Table IV. are fo felected that each pair is equally
affected by aberration, or nearly fo; it will not be neceffary to
apply thefe two corrections.

quantity of the deviation may be found as
follows:

Add the declinations of the two ftars
together, (one being north and the other
fouth) and the fum will be the difference
of their declinations.

Take the difference between the obferved difference of the right afcenfions of
the two ftars, and the apparent difference
of their right afcenfions; to the logarithm
of which (in feconds) add the log. co-fines
of their declinations, the log. co-fecant of
the difference of their declinations, and the
log. fecant of the latitude of the place of
obfervation : their fum, rejecting 40 in the
index, will be the logarithm of the number
of feconds *in time*, which expreffes the angle
made by the inftrument and the meridian.
This being done, the inftrument may be
brought into the plane of the meridian by
the inftructions given in Art. 53.

EXAMPLE.

Suppofe that at Barbadoes, latitude
13° 5′ N. the ftar Fomalhaut, paffed a tranfit inftrument at 10ʰ 24′ 12″,4 by a watch,
on the 28th of September, 1796, and that
at 10ʰ 32′ 1″,8 α Pegafi paffed the fame inftrument: fuppofe alfo, that on the 29th,

Fomalhaut paffed the fame inftrument at
10ʰ 20′ 22″ and α Pegafi at 10ʰ 28′ 11″,2:
How much did that inftrument deviate from
the meridian, and which way?

Æ of Fomal. Tab. V. 22ʰ 46′ 16″,6 of β Peg. 22ʰ 53′ 51″, 2
Annual variation .. +   3, 32 ......   2, 87
Three fourths dᵒ. ...   2, 48 ......   2, 16

Mean Æ Sept. 1796, 22 46 22, 4 ..... 22 53 56, 2
Deviation (Art. 55.)   — 1 ..... .   + 1

      22 46 22, 3 ..., . 22 53 56, 3
      22 46 22, 3

Apparent difference of right-afcenfion   0 7 34, 0

      24ʰ   24ʰ
Fom. paff. on the 29th 10 20′ 22″,0 β Peg. at 10 28 11, 2

      34 20 22, 0 ...... 34 28 11, 2
Fom. paff. on the 28th 10 24 12, 4 ...... 10 32 1, 8

Time of a revolution 23 56 9, 6 ..,... 23 56 9, 4
The m. of the two is 23 56 9, 5
Fomalhaut paffed before β Pegafi on the 28th 7′ 49″,4
      on the 29th 7 49 ,2

      Mean obferved difference 7 49 ,3

As 23ʰ 56′ 9″,5 is to 24ʰ fo is 7′ 49″,3
to 7′ 50″,55, the obferved difference of right
afcenfion : the difference between this and
7′ 34″,0 is 16″,55, the quantity by which the
obferved difference of right afcenfion ex-
ceeds the apparent; and the ftar neareft
the elevated pole paffed laft: confequently,
that end of the telefcope which is directed

toward the elevated pole lay weft of the meridian.

58. To bring the inftrument into, or nearer the meridian, let us fuppofe a ftar at the equator, as before, was 36″¼ paffing from one wire to another. By Table IV. the declination of Fomalhaut, is 30° 41′ 36″ S. and the declination of β Pegafi 26° 58′ 54″ N. their fum, which is the difference of their declinations, is 57° 40′ 30″.

Error in Æ		16″, 55	Log. 1, 21880
Fomalhaut's declin.	30° 42′	co-fine	9, 93442
β Pegafi's declin.	26 59	co-fine	9, 94995
Different declin.	57 41	co-fec.	10, 07309
Lat. Barbadoes	13 5	fecant	10, 01142
Error of the inftru.		15″, 4	Log. 1, 18768

This is very near three-feventh parts of the interval between the wires in the focus of the telefcope; and therefore, having found an object in the horizon which is cut by one of the wires, turn the fcrew *g* until that object appears to be removed three-feventh parts of the fpace, which is between the wire it was on before, and the next to it (apparently) toward your right hand, and the inftrument will be in, or very near the meridian ; and its real pofition may be examined by the fame, or any other pair of ſtars, and adjuſted till you find it exactly in the meridian.

TO OBSERVE THE SUN'S TRANSIT OVER THE MERIDIAN.

59. From the latitude of the place and the ſun's declination find its meridional altitude by Note 1. Art. 48; and, a few minutes before noon, ſet the inſtrument to this altitude, by means of the ſemi-circle L Z I and index E K. Apply your eye (defended by a dark glaſs) to the teleſcope, and wait till you ſee the firſt limb of the ſun enter it; which will be (apparently) on the weſt ſide, becauſe the teleſcope inverts objects. When this happens let your aſſiſtant attend to the watch; and, when the firſt limb of the ſun touches the firſt wire, bid him *mark* the ſecond, and parts of a ſecond, which are ſhewn by the watch; and which he muſt ſet down, in the firſt column of a paper that contains five, ready ruled for the purpoſe. He muſt then prefix the minute, and attend again to the watch. When the ſun's firſt limb arrives at the ſecond wire, bid him again *mark* the ſecond, and parts of a ſecond ſhewn by it; which being ſet down in the ſecond column of his

paper, and the minute prefixed, he muſt attend again to the watch. And in this manner the times when the ſun's firſt limb arrives at every one of the wires muſt be obſerved, and noted down in its proper column. The times when the ſecond limb arrives at each of the five wires muſt be obſerved in the ſame manner, and written in their proper columns, under thoſe of the firſt. If the wires in the focus of the te- leſcope be ſo diſpoſed, that there is not time to obſerve the firſt limb at all the five wires before the ſecond limb arrives at the firſt wire, the obſervation of the firſt limb at the fifth wire muſt be omitted, and, in this caſe, the obſervation of the ſecond limb at the firſt wire may be omitted alſo, as it will be of no uſe.

60. The mean of the times when the two limbs of the ſun, were at the middle wire will be the time of apparent noon by the watch; and, if the wires are equi- diſtant, (as they ought to be) the mean of the two times when the firſt limb was at the firſt wire, and the latter limb at the fifth wire will alſo be the time of noon. Alſo the mean of the two times when the firſt limb was at the ſecond wire, and when the latter limb was at the fourth wire will

be the time of noon; as alfo the mean of
the times when the firft limb was at the
fourth wire and the latter limb at the fecond. If the firft limb was obferved at the
laft wire, and the latter limb at the firft,
the mean of thefe two obfervations will alfo
be the time of apparent noon; and the
mean of all thefe refults, if they differ, as
they moft likely will, is the time of apparent noon by the watch.

TO FIND THE RATE, WHICH A WATCH GOES AT BY OBSERVATIONS OF THE SUN'S TRANSIT OVER THE MERIDIAN.

61. Obferve the time when the fun
tranfits the meridian of the place, by
Art. 59, every day at noon, or as often as
opportunities offer. The equation of time
muft then be taken from the Nautical Almanac by Art. 24; and, if it be marked additive, it will be the fame as the time by the
watch when the fun's center was obferved
to pafs the meridian, if the watch be right.
If they differ, that difference is what the
watch is too faft, or too flow for mean
time : and it is too faft if the time by the
watch be greater than the equation of time;
and too flow if the time by the watch be
lefs. If the equation of time be fubtrac-

I

tive, take it from twenty-four hours, com-
pare the obferved time when the fun's center was on the meridian with the remainder, and the difference between them will
be what the watch is too faft, or too flow,
according as the time by the watch is the
greater or the lefs. Thefe obfervations,
when the voyage is expected to be of a
confiderable length, ought to be continued
for a month, at leaft: indeed, the longer
they are continued, in all cafes, the better;
but in this the obferver muft be governed
by circumftances. They muft always, however, be continued as near as poffible to the
time when the fhip is expected to fail, that
there may be as little chance as poffible left
for the watch to alter its rate of going after
the obfervations are clofed.

62. The times by the watch when the
fun's center was obferved on the meridian,
muft be written one under another, againft
the days of the month when they were obferved; and it is the day that began at the
inftant when the fun's center was on the
meridian which is to be fet before the obferved time; and not the day which ended
then, as is the cuftom with feamen. The
equation of time, or its fupplement to 24
hours, according as it is additive, or fub-

tractive, muſt be ſet in another column, againſt the obſerved times of noon, and the difference between them in a third, with the ſign + or —, according as the watch is too faſt or too ſlow for mean time. This is all that is neceſſary to be done till all the obſervations are made.

63. When the ſhip is ready to ſail, add a fourth column to your paper, take the difference between what the watch was too faſt or too ſlow on the firſt day of obſervation, and what it was too faſt or too ſlow on the ſecond, and put it in the fourth column, oppoſite the ſpace which is between the two numbers of which it is the difference. Take alſo the difference between what the watch was too faſt or too ſlow on the ſecond day, and what it was too faſt or too ſlow on the third; between what it was too faſt or too ſlow on the third, and what it was too faſt or too ſlow on the fourth; and ſo on. Place theſe differences alſo in the fourth column, oppoſite the ſpaces which are between the two numbers of which they are, reſpectively, the difference. Thoſe differences will be the gain or loſs of the watch in the twenty-four hours, which they reſpectively ſtand againſt: And it muſt be obſerved, that the

watch is gaining if it be too faft for mean time, and the numbers in the third column increafe; or if it be too flow, and the numbers in the third column decreafe: but the watch is lofing if it be too faft for mean time, and the numbers in the third column decreafe; or if it be too flow, and the numbers in the third column increafe.

. 64. By making daily obfervations in the manner here recommended, it will be feen whether the watch alters its rate of going while it is under trial, which is abfolutely neceffary to be known; becaufe, if it does, all thofe obfervations muft be rejected which were made before the alteration happened, and thofe only retained which were made afterward.

65. If no material alteration happened in the rate of the watch's going, during the time of trial, take the difference between what the watch was too faft or too flow on the firft day of obfervation, and what it was too faft or too flow on the laft, if they be of the fame kind; that is, both too faft, or both too flow; but add them together if the watch was too faft in one inftance and too flow in the other; this difference, or fum, muft be divided by the number of days which elapfed between the firft and

laſt day's obſervations, and the quotient will be the number of ſeconds and decimal parts that the watch gains or loſes in a day. And it is manifeſt, that if the watch be faſter at the end of the trial, than it was at the beginning, it is gaining; and if it be ſlower, it is loſing.

66. If any conſiderable alteration happened in the rate which the watch went at, inſtead of taking the difference between what the watch was too faſt, or too ſlow, on the firſt and laſt days, take the difference between what the watch was too faſt or too ſlow on the day after that, when the alteration in its rate happened, and what it was too faſt or too ſlow on the day when the laſt obſervation was made, and divide by the number of days which elapſed between them. The following example will make this very plain.

67. Suppoſe the obſerved times when the ſun's center paſſed the meridian of Barbadoes, in the month of December 1793, were as follow; what was the loſs or gain of the watch on mean time?

1793.	Obf. Times of ☉'s Tran.			Mean Time of app. Noon.			Watch too faft.				Daily gain.
	H.	M.	s.	H.	M.	s.		H.	M.	s.	s.
☉ Decem. 1	3	50	34,0	23	49	41,7	+	4	0	52,3	
☽ - - - 2	3	51	2,1	23	50	5,4	+	4	0	56,7	+ 4,4
♂ - - - 3	3	51	30,6	23	50	29,7	+	4	1	0,9	+ 4,2
☿ - - - 4	3	51	59,2	23	50	54,6	+	4	1	4,6	+ 3,7
♃ - - - 5	3	52	26,2	23	51	20,1	+	4	1	6,1	+ 1,5
											+ 1,8
♀ - - - 6	3	52	54,0	23	51	46,1	+	4	1	7,9	+ 1,4
♄ - - - 7	3	53	21,9	23	52	12,6	+	4	1	9,3	+ 1,5
☉ - - - 8	3	53	50,3	23	52	39,5	+	4	1	10,8	+ 1,3
☽ - - - 9	3	54	18,9	23	53	6,9	+	4	1	12,0	+ 1,1
♂ - - - 10	3	54	47,7	23	53	34,6	+	4	1	13,1	+ 0,7
☿ - - - 11	3	55	16,4	23	54	2,6	+	4	1	13,8	
♃ - - - 12	3	55	46,7	23	54	31,0	+	4	1	15,7	+ 1,9
♄ - - - 14	3	56	48,7	23	55	28,6	+	4	1	20,1	+ 2,2
☉ - - - 15	3	57	19,8	23	55	57,7	+	4	1	22,1	+ 2,Q
☽ - - - 16	3	57	51,3	23	56	27,0	+	4	1	24,3	+ 2,2
											+ 1,4
♂ - - - 17	3	58	22,2	23	56	56,5	+	4	1	25,7	+ 1,5
☿ - - - 18	3	58	53,4	23	57	26,2	+	4	1	27,2	+ 1,3
♃ - - - 19	3	59	24,6	23	57	56,1	+	4	1	28,5	+ 1,8
♀ - - - 20	3	59	56,3	23	58	26,0	+	4	1	30,3	+ 1,6
♄ - - - 21	4	0	27,9	23	58	56,0	+	4	1	31,9	+ 1,8
♂ - - - 24	4	2	3,4	0	0	26,1	+	4	1	37,3	
☿ - - - 25	4	2	33,6	0	0	56,0	+	4	1	37,6	+ 0,3
♄ - - - 26	4	3	5,0	0	1	25,8	+	4	1	39,2	+ 1,6
♀ - - - 27	4	3	36,1	0	1	55,4	+	4	1	40,7	+ 1,5
♄ - - - 28	4	4	08,0	0	2	24,9	+	4	1	43,1	+ 2,4
											+ 2,9
☉ - - - 29	4	4	40,2	0	2	54,2	+	4	1	46,0	+ 2,2
☽ - - - 30	4	5	11,5	0	3	23,3	+	4	1	48,2	+ 1,5
♂ - - - 31	4	5	41,9	0	3	52,2	+	4	1	49,7	

68. Here it appears that the watch went confiderably fafter the three firft days than it did afterward; I therefore reject thefe three days, and take the difference between 4ʰ 1′ 4″,6, what the watch was too faft on the fourth, and 4ʰ 1′ 49″,7, what it was too faft on the 31ft, and find it 45″,1, which I divide by 27, the number of days elapfed, and the quotient, 1″,6704, is the quantity which the watch gained on mean time in one day. It may be proper to repeat that the obfervations on the intermediate days are abfolutely neceffary, though no ufe is made of them in deducing the rate of the watch's going; becaufe, without them, it would be impoffible to tell whether the watch had altered its rate of going while under trial, or not; and, of courfe, whether it would be proper to make ufe of the whole time, or only a part of it; and what part.

69. But notwithftanding the fimplicity of the calculations, and the eafe with which the obfervations are made, render this method of finding the rate of a watch vaftly convenient in fixed obfervatories, it is not fo well adapted to the fkill and opportunities of feafaring men, in general, as fome others are; for it requires not only a con-

fiderable degree of knowledge in practical aftronomy, but fome time alfo, to get a tranfit inftrument into the plane of the meridian. And if the inftrument be not pretty exactly in the meridian, the obferver will not only get the abfolute quantity of time, which the watch is too faft, or too flow, wrong; but will, if there be any confiderable change in the fun's declination, while the watch is under trial, determine the rate of its going erroneoufly alfo. On this account it will be better to find the rate of the watch, by obferving the tranfits of a fixed ftar; and the abfolute time by obfervations of the fun's altitude, taken with an Hadley's quadrant, from an artificial horizon. For, when a fixed ftar is ufed, it is not neceffary that the inftrument be even near the meridian; though it will be convenient to place it as near it as a well-defined mark can be found, to adjuft it to: and, therefore, the inftrument may be fet up, and adjufted ready for obfervations of this kind, in the fpace of an hour. Moreover, by this method, the computations are ftill more fimple than they are when the fun is made ufe of; for, having taken the times by the watch when any fixed ftar tranfits the vertical circle which the inftrument moves in, every day, or as often as an opportunity offered, and for such a length of time as may be thought neceſſary; put theſe times one under another, againſt the days of the month they were obſerved on. Take the difference between the firſt and ſecond, between the ſecond and third, and ſo on; and take alſo, the difference between each of theſe differences and 3′ 56″,55:* and theſe laſt differences will be the gain or loſs of the watch on mean time, in 24 hours. And the watch gains, if the firſt mentioned differences are leſs than 3′ 56″,55; but it loſes if they are greater. If the obſervations have been made regularly every day, and the watch has conſtantly loſt, or conſtantly gained on mean time, add the ſeveral gains or loſſes into one ſum, and divide it by the number

* The fixed ſtars having no apparent annual motion, re-turn to the meridian exactly in the time that the earth makes a revolution on its axis ; while the ſun, having been carried, during the earth's revolution on its axis, ſome diſtance eaſt-ward, by its apparent annual motion in the ecliptic, will not return to the meridian until the earth, beſide making a com-plete revolution on its axis, has followed it through an arc which is equal to its diurnal motion in the ecliptic. And as this arc, on a medium, is equal to 59′ 8″,3, according to the beſt obſervtions, the mean ſolar day will be longer than the fiderial day by 3′ 56″,55, which is the time that the earth takes to revolve through an arc of 59′ 8″,3.

K

of days the watch has been under trial, and the quotient will be the real gain or lofs which it makes in one day.

70. If the obfervations have been repeated every day, and, on comparing them, the watch be found to have gained fome days, and loft others; take the fum of all the gains, and the fum of all the loffes, feparately, fubtract the lefs fum from the greater, and divide the remainder by the number of days which the watch was under trial, and the quotient will be the real gain or lofs of the watch in one day, according as the fum of the gains, or loffes, is greateft.

71. If, during the trial, the obfervations be omitted on fome days, fo that two or more days intervene; $3' 56'',55$ muft be multiplied by the number of days which elapfe between the obfervations, and the difference between the times of the ftar's tranfit muft be compared with the product: the difference between them will be the gain or lofs of the watch in the interval between the obfervations; and, if this be divided by the number of days which have intervened, the quotient will be the gain or lofs in one day: but it is the gain or lofs in the whole interval which muft be

taken in finding the real gain or lofs of the
watch, at the end of the trial. The following example will explain the whole pro-
cefs fully.

72. Suppofe the times by a watch
when the ftar Aldebaran paffed a tranfit
inftrument placed nearly in the meridian
of Madras, were as follow: it is required
to find how much the watch gained or loft
on mean time?

1794.	Obf. Times of the *'s Tranf.	Diffe-rence.	Diff. bet. m. Sol. & fid. Day,	Watch gains, or lofes on mean T.	Remarks
	H. M. s.	M. s.	M. s.	s.	
♀ Jan. - 3	9 22 17,42	3 56,69	3 56,55	− 0, 14	
♄ - 4	9 18 20,73	3 56,37	3 56,55	+ 0, 18	
☉ - 5	9 14 24,36	3 56,31	3 56,55	+ 0, 24	
☽ - 6	9 10 28,05	3 56,57	3 56,55	− 0, 02	
♂ - 7	9 6 31,48	3 57,61	3 56,55	− 1, 06	
☿ - 8	9 2 33,87	3 57,34	3 56,55	− 0, 79	
♃ - 9	8 58 36,53	7 51,46	7 53,10	+ 1, 64	In 2 days
♄ - 11	8 50 45,07	3 55,61	3 56,55	+ 0, 94	
☉ - 12	8 46 49,46	11 49,29	11 49,65	+ 0, 36	In 3 days
☿ - 15	8 35 0,17	3 55,55	3 56,55	+ 1, 00	
♃ - 16	8 31 4,62	3 56,41	3 56,55	+ 0, 14	
♀ - 17	8 27 8,21				

73. Here it may be obferved that the fum of all the watch's gainings is 4″,50, and the fum of all its lofings is 2″,01 : the difference between them is 2″,49; which being divided by 14, the number of days the watch was under trial, will give 0″,178 for the daily gain of the watch.

74. The rate which a watch goes at is obtained this way with much lefs trouble than by any other; but the abfolute time is not given by it; nor, of courfe, how much the watch is too faft, or too flow, for mean time at the meridian it is tried under. This muft be found by obferving altitudes of the fun, immediately before the fhip fails, from the horizon of the fea; or double altitudes of the fun, from an artificial horizon, if the former be not open toward the eaft or weft at the place of obfervation.

E X A M P L E.

Admit that on the 17th, the laft day on which the ftar was obferved, the following obfervations were made to find how much the watch was too faft or too flow for mean time; the height of the obferver's eye above the fea being 27 feet.

Times by the watch.	Alt. of the Sun's L. L.	Appa. Time at Madras 19ʰ 23′
		Long. of Madras in T. 5 22
19ʰ 7′ 53″	10° 9′ 0″	Appa. Time at Green. 13 46
8 34	16 15	
9 17	24 45	
9 58	32 30	Sun's femidiameter . . . . 16′ 18″
10 42	41 00	Sun's horizontal parallax 9
11 22	48 45	
		16 27
6)57 46	6)172 15	Dip of the hor. 4′ 57½″ } 10 0½
		Refraction 5 3
19 9 37,7	10 28 42½	
	+ 6 26½	Correct. of the fun's alt. 6 26½
	10 35 9	True altitude.

Sun's dec. the 17th at noon Green. 20° 38′ 22″ S.
Correction for 13ʰ 46′ fubtract   7 4

Sun's corrected declination . . . . . 20 31 18 S.
                              90°
Ship's latitude 13 4′ 54″ N. fecant 10. 01142

Co-latitude   76 55 6
Sun's declin. 20 31 18 S. fecant 10. 02847

Meridional alt. 56 23 48 N.S. 83289 } 64918 log. 4. 81237
Sun's true alt. 10 35 9 N.S. 18371 }

Time from noon . . . . 4ʰ 52′ 57″,6 log-rifing 4. 85226
                   .   24

Apparent time 19 7 2,4
Equation of time add   10 54, 9

Mean time 19 17 57, 3
Time by the watch   19 9 37, 7

Watch too flow for m. T. 8 19,6

    More fets of altitudes may be taken if there be oppor-
tunities for it; by which means the deviation of the watch
from mean time will be obtained with greater certainty.

75. As some, however, may not wish to put themselves to the expence of, and others not to be encumbered with an additional instrument; and more still not to give themselves the trouble of learning the use of it; I shall now shew how the rate of a watch may be found, with equal accuracy, by Hadley's sextant and an artificial horizon.

TO FIND THE RATE WHICH A WATCH GOES AT BY EQUAL ALTITUDES OF THE SUN.

76. In the morning, when the sun is, at least, four points from the meridian; but the nearer to the east the better, place your artificial horizon * in a convenient

---

* There are different contrivances for this purpose; but an oblong trough, filled with quickfilver, and sheltered from the wind, is the only one that can be depended on. I have tried them all; and find that the circular one, with a bubble in the center, though, to appearance, perfect when first made, soon grows useless; the glass which covers the fluid altering its figure, I suppose, from the pressure of the fluid against it. Those which are formed by floating pieces of glass on quickfilver, though filled with the utmost care, are equally unsafe: for the instant the glass comes near any side of the vessel it floats in, it loses it horizontality, and will not recover it: the quickfilver must therefore be taken out, and put in again: but the trouble attending this operation is the least part of the evil, for when the sun is either rising or falling fast, the observer will not readily discover whether any alteration of this kind has happened or not. The best shelter for the quickfilver is, undoubtedly,

fituation: take a Hadley's fextant, fcrew
the telefcope into its place, and bring the
dark glaffes to intervene, on each fide the
horizon glafs. Having found that fituation
in which you can fee the image of the fun,
in the quickfilver, conveniently, bring the
inftrument to your eye, and find the fame
image in the telefcope : then move the index of the fextant until you bring the
image, reflected from the index glafs, into
the telefcope alfo ; and having made it pafs
over, and a little way beyond the image on
the quickfilver, fix the index by the fcrew-
clamp very fecurely. Place a perfon at the
watch, and obferve carefully when the two
neareft limbs of the two images come exactly into contact, by the increafing altitude of the fun ; and, the inftant this hap-

a roof formed by two plates of glafs ; the two fides of each
being ground perfectly plane, and parallel to one another : but,
to prevent any bad effects from their being otherwife, care muft
be taken to put the roof on the fame way ; that is, fo that the
fame fide may be toward the obferver in the afternoon that was
toward him in the morning. The late Mr. Reuben Burrow-
who, whatever faults he might have, was certainly a very ingenious man, has recommended a piece of mufquetto curtain,
ftretched in a frame, as a good fhelter for the quickfilver.
See Afiatic Refearches, Vol. I. I have tried this mode of fhel-
tering the quickfilver, and found it anfwer the purpofe very
well, when the wind was not very ftrong, and the fun was reafonably bright ; the fun's limb, however, is not fo diftinct as
it is when the quickfilver is fheltered by a glafs roof.

pens, bid your affiftant *mark* the fecond,
and parts of a fecond, fhewn by the watch
when it happens. Thefe being written
down, and the proper hour and minute annexed to them, bid him attend again to the
watch, while you look out for the contact
of the other limbs of the two images,
which, by this time, will be got one on to
the other, and approaching to a co-inci-
dence. They will afterward begin to fe-
parate; and when their limbs are again
exactly in contact, bid your affiftant *mark*
the fecond, and parts of a fecond when it
happened: put them down, and annex the
proper hour and minute. The fextant muft
now be put carefully away, without alteration in any refpect, in a place where it
will not be liable to be difturbed; and another fextant, or as many of them as are
at hand, may be taken one after another,
the fame obfervations made with them, and
the fextants put away with the fame care
till the afternoon.

77. Eftimate as near as you can, the
time from noon when the laft obfervation
was made, allowing for the equation of
time; and a little before that time in the
afternoon fet out your horizon, take the
fextant which the laft obfervation was

made with in the morning, direct the telefcope to the fun's image in the quickfilver,
and if it be near the time of obfervation,
the image from the index glafs will be in
the telefcope alfo; and a little attention
will fhew you that the images approach
each other. Watch carefully till their
neareft limbs come into contact, and the
inftant they do bid your affiftant *mark* the
fecond, and parts of a fecond fhewn by the
watch. In the fame manner mark the fecond and parts of a fecond, when the images having paffed over each other, their
limbs are again in contact at their fe   para-
tion; which, as well as the fecond and
parts, when the former contact happened,
with their proper hours and minutes annexed muft be written down, oppofite the
obfervations made with the fame fextant in
the morning. And the firft obfervation
in the afternoon muft always be written
oppofite the latter morning obfervation;
and the latter evening obfervation oppofite the firft morning obfervation.

79. Repeat thefe obfervations with each
of the fextants which were ufed in the
morning, fetting the obfervations oppofite
the obfervations made with the fame fextant in the morning; the firft in the after-

noon againſt the latter morning one as before; and you will have as many ſets of
obſervations, two in each ſet, as you uſed
ſextants.

80. When you fix the index to the
arch by the clamp, before the morning obſervation, it will not be amiſs if you ſet the
index very exactly, to ſome even diviſion
by means of the tangent ſcrew; and note
what diviſion it was ſet to againſt the obſervation made with that ſextant. When
this precaution is taken, you may examine,
before you begin to obſerve in the afternoon, whether the index has been altered
by any accident or not, and if it has, it
may be placed to the ſame diviſion again.
But if it be found that any one of the ſextants has been altered, the obſervations
made with that ſextant ſhould be uſed
with caution; and rejected if they be found
to differ from thoſe made with the others,
notwithſtanding the index of ſuch ſextant
may have been reſtored to the proper diviſion before the afternoon obſervation.

81. The moſt proper time for making
theſe obſervations, is when the ſun is nearly
due eaſt and weſt. When, therefore, ſeveral obſervations are to be taken, it will
be prudent to begin before the ſun comes

to the eaſt point of the horizon, in the
morning, when it can be done; but not
before ſix o'clock, nor before the ſun is
8 or 10 degrees high : indeed, too implicit
a confidence muſt not be placed in obſervations made at theſe altitudes, if there
be any material difference between the
heights of the thermometer at the times of
the morning and evening obſervations.

82. In order to deduce the time of
noon from the obſervations, and by that
means determine how much the watch is
too faſt or too ſlow for mean time, add 24
hours to the time of the laſt obſervation in
the afternoon, and take the time of the
firſt obſervation in the morning from it,
the remainder will be the interval between
the obſervations. Take the half of this
interval, and add it to the time of the
morning obſervation, and it will give the
time of apparent noon nearly. Take the
ſun's longitude for noon at the place of obſervation, from page II. of the Nautical
Almanac; with which, the latitude of the
place, and the half-interval between the
obſervations, take the equation to equal
altitudes from the tables at the end of this
work; and add it to, or ſubtract it from the
time of noon, nearly, as the ſigns annexed

to the fun's longitude direct, and it will give the time of apparent noon by the watch, according to this pair of obferva- tions. Repeat the operation with each pair of obfervations which may have been made, and you will have as many times of appa- rent noon by the watch as there are pairs of obfervations, which ought all to agree exactly; but as that can feldom be expect- ed, and as there will generally be fome fmall differences between them, the mean of them all muft be taken for the time of apparent noon by the watch.

EXAMPLE.

Admit that on the 25th of Auguft, 1793, the following obfervations of equal altitudes were made at Quebec.

Ther-mome-ter.	Morning.			Afternoon.			Dou. Alt.		Ther-mome-ter.		
	H.	M.	s.	H.	M.	s.	°	′			
56° {	19	26	53, 9	4	35	43, 3	} 45	00	67°	U.	L.
	19	30	2, 3	4	32	35, 7				L.	L.
58 {	20	4	25, 6	3	58	14, 6	} 57	30	68	U.	L.
	20	7	36, 0	3	55	5, 0				L.	L.

Upp. Limbs	Low. Limbs	Upp. Limbs	Low. Limbs	
H. M. S.	H. M. S.	H. M. S.	H. M. S.	
28 35 43,3	28 32 35,7	27 58 14,6	27 55 5,0	Aftern. obfer.
19 26 53,9	19 30 2,3	20 4 25,6	20 7 36,0	Morn. obfer.
9 8 49,4	9 2 33,4	7 53 49,0	7 47 29,0	Interval.
4 34 24,7	4 31 16,7	3 56 54,5	3 53 44,5	Half interval.
0 1 18,6	0 1 19,0	0 1 20,1	0 1 20,5	Noon nearly.
+ 18,3	+ 18,2	+ 17,1	+ 17,0	Equa. Ta. I.
− 1,1	− 1,2	− 1,5	− 1,5	Equa. Ta. II.
0 1 35,8	0 1 36,0	0 1 35,7	0 1 36,0	True time of
			35,7	noon by the
			36,0	watch.
			35,8	
			4)143,5	

Time of noon by the watch .... 0ʰ 1′ 35″,9

Mean time of appa. noon (Naut. Alm. p. II.) 0 1 35, 3

Watch too faft for mean time 00, 6

And in this manner it may be found how much the watch is too faft or too flow for mean time, every day at noon when equal altitudes of the fun can be obtained; and, from thence, whether the watch gains or lofes, and how much, as in Art. 68.

PROBLEM VI.

83. To find the longitude at fea by a Time-keeper.

RULE.

Obferve the altitude of the fun's limb, either in the morning or evening, when it is, at least, three points of the compass from the meridian, and note the time when it was obferved by the Time-keeper.

Multiply the daily rate of the watch by the number of days, which have elapfed fince that on which the laft obfervation was made for finding it, and add the product to the time fhewn by the watch when the fun's altitude was obferved, if the watch be lofing, but fubtract it from that time if the watch be gaining. To the fum, or remainder, add what the watch was too flow, or fubtract from it what the watch was too faft for mean time at the place where its rate was found, on the day when the laft obfervation was made for finding it, and the refult will be the mean time at that place when the fun's altitude was obferved. To this time add the longitude of the place in time where the rate of the watch was found, if it be weft; or fubtract the longitude in time from it, if it be eaft, and the fum or remainder will be the mean time at Greenwich.

To this time find the fun's declination by Problem III. and correct the obferved altitude of the fun's limb for the dip of the horizon, refraction, parallax, and femidiameter; with which, the latitude of the ſhip, and the ſun's declination, find the mean time at the ſhip, by Problem IV.

Take the difference between the mean time at the ſhip, and the mean time at Greenwich, and it will be the longitude of the ſhip in time; eaſt, if the time at the ſhip be greater than the time at Greenwich, but weſt if it be leſs.

EXAMPLE I.

After having found the rate of a watch to be gaining 1″,67, as in Art. 68, and that it was too faſt for mean time at Barbadoes, on the 31ſt of December, 1793, by 4^h 1′ 49″,7; let us ſuppoſe that on the 4th of February, 1794, in the afternoon, latitude 44° 26′ N. the following obſervations were taken: what was the longitude of the ſhip; the height of the obſerver's eye above the ſurface of the ſea being 21 feet?

		H. M. S.
Time by the Watch	Alt. of the ☉'s L. L.	T.-keeper too faſt Dec, 31, 1793, 4 1 49,7
		Gain to Feb. 4, 1794 = 1″, 67×35ᵈ. + 58,5
		Time-keeper too faſt Feb. 4, 4 2 48,2
5ʰ 2′ 51″	9° 17′ 15″	
3 44	9 8 45	Sun's ſemi-diameter .... 16′ 16″
4 40	8 59 30	Sun's horizontal parallax 9
5 49	8 50 00	
		16 25
4)17 4	36 15 30	Dip of the horizon 4′ 22″ } 10 10
		Refraction .... 5 48 }
5 4 16	9 3 52	
4 2 48	+ 6 15	Correction of the ſun's altitude 6 15
1 1 28	9 10 7	True altitude.
3 58 45		Longitude of Barbadoes, W.
5 0 13		Mean time at Greenwich,

Sun's declination for noon at Greenwich 16° 3′ 24″ S.

Correction for time at Greenwich ..... − 3 31

Sun's correct declination .... 15 59 53 S.

90° 00′

Ship's latitude 44 26 N. ſecant 10. 14626

Co-latitude 45 34
Sun's declin. 16 00 S. ſecant 10. 01715

Merid. altit. 29 34 N. S. 49344 }
Sun's obf. alt. 9 10 N. S. 15931 } 33413 Log. 4. 52393

Appa. time at the ſhip 3ʰ 56′ 29″ Log-riſing 4. 68734
Equation of time, add 14 26

Mean time at the ſhip 4 10 55
Mean time at Green. 5 0 13

Longitude in time 0 49 18 = 12° 19½ W.

EXAMPLE II.

March 29th, 1794, latitude 55° 9'¼ N. the following obfervations were made to determine the longitude by the fame watch.

			H. M. S.
Time by the Watch	Alt. of the ⊙'s L. L.	T.-keeper too faft Dec. 31, 1793,	4 1 49,7
		GaintoMar.29,1794=1″,67×88¼ᵈ.	+ 2 27,8
18ʰ 8' 30″	11°27' 15″	Time-keeper too faft Mar. 29,	4 4 17,5
9 28	34 45		
10 26	42 00	Sun's femi-diameter	16' 3″
11 27	49 30	Sun's parallax in altitude	9
12 32	58 00		
			16 12
5)52 23	5)211 30	Dip of the horizon 4' 22″	} − 8 53
		Refraction 4 31	
18 10 28,6	11 42 18		
4 4 17,5	+ 7 19	Correction of the fun's altitude	7 19
14 6 11	11 49 37	True altitude.	
3 58 45	....	Longitude of Barbadoes, W.	
18 4 56	....	Mean time at Greenwich.	

Sun's declination for noon at Greenwich   3° 58' 00″ N.
Correction for time at Greenwich .....   — 5 49

Sun's correct declination ....   3 52 11 N.

90° 00'
Ship's latitude 55 9¼ N. fecant 10. 24313

Co-latitude 34 50½
Sun's declin. 3 52½ N. fecant 10. 00099

Merid. altit. 38 42⅔ Nat. S. 62539 } 42043 Log. 4. 62369
Sun's obf. alt. 11 49¼ Nat. S. 20496 }

4ʰ 59' 9″   Log-rifing 4. 86781
24

Appa. time at the fhip 19 0 51
Equation of time   + 4 32

Mean time at the fhip 19 5 23
Mean time at Green. 18 4 56

Longitude in time   1 0 27 = 15° 6'¾ Eaft.

M

EXPLANATION and USE

OF THE

TABLES of EQUATIONS

TO

EQUAL ALTITUDES.

IN the Nautical Almanac for 1773, the Commiffioners of the Board of Longitude, were pleafed to publifh a fet of " Tables of Equations to Equal Altitudes," which I had computed, principally for amufement, during the many dreary hours I paffed on the coaft of Hudfon's Bay in 1768, and 1769. That Almanac has long been out of print, and of courfe the tables are not to be procured; which has induced many gentlemen who are employed in, or amufe themfelves with practical aftronomy, to wifh I would reprint them. In computing the tables which were publifhed in the Nautical Almanac I made ufe of interpolations; and, before I reprinted them, I was defirous of knowing how far thefe interpolations might be depended on: I therefore employed fuch of the Boys who are under my care at Chrift's

Hofpital, as I found capable, in recomput-
ing them, during their leifure hours; and
the tables which I now offer to the public,
have been entirely recomputed with the
utmoft rigour: Every computation was
made by two Boys, feparately, and at diftant times; and the whole has been compared, and revifed where any difference appeared, by myfelf. I have alfo, now, given
the equations in feconds and decimal parts;
which renders them more convenient than
they were before; and, in fome degree, more
accurate.

In calculating thefe Tables I adhered to
the formula which I made ufe of before;
namely, the change in the fun's delination,
(during half the interval between the obfervations) multiplied by the co-fecant of
the meafure of that half-interval, multiplied again by the tangent of the latitude
of the place of obfervation, ± the faid
change in the fun's declination, multiplied
by the tangent of the fun's declination,
and multiplied again by the co-tangent of
the meafure of the half-interval*. The

* It is thus inveftigated. Let P Z C, Fig. 6, reprefent
the meridian of the place, P the elevated pole; Z the zenith,
A C F the almicanther the fun is on at the time of obfervation,
B the fun's place at the morning obfervation. F its place
when the obfervation was made in the afternoon, and AGD

numbers in Table I. are formed by multiplying the change in the fun's declination

the parallel of declination which the fun was on at the middle of the interval between the obfervations, and which cuts the almicanther in A and H. Defcribe the hour circles P A, P B, P H, P F; and bifect the angle B P F, which is the meafure of the interval between the obfervations, with the hour-circle P K G, and it is manifeft that the angle B P G is the meafure of half the interval between the obfervations; which being added to the time fhewn by the clock or watch when the morning obfervation was made, will give the time when the fun was on the hour-circle P K G: but the fun was on the meridian P Z C at the inftant of apparent noon; confequently, the meafure of the angle Z P G is the equation, which muft be applied to the middle time between the two obfervations, to give the time of apparent noon by the clock or watch. It is moreover manifeft from the figure itfelf that when the fun is in the afcending figns; that is, while it is approaching the elevated pole, the time of apparent noon precedes the middle time between the obfervations; and, confequently, the equation muft be fubtracted from the middle time to give the time of apparent noon: and when the fun is receding from the elevated pole, the equation muft be added to the middle time to give the time of apparent noon.

As the variation of the fun's declination in the firft half-interval is never fenfibly different from its variation in the latter, the angle A P B will, evidently, be equal to the angle H P F; and, confequently, each of them is equal to Z P G, the equation fought: make P K = P Z, and defcribe the great circles K A and K B: then, in the fpherical triangle B K P, the two fides B K, K P, may be confidered as being conftant, while the fide P B, which is the complement of the fun's declination varies; and it is evident, that while P B varies the quantity B I the angle Z P B is altered by the quantity B P A, which has been fhewn to be equal to the equation fought. Now, by Art. 256, of Simpfon's Fluxions, or Theorem 23 of Cotes *de Eftimatione Errorum in mixtâ Mathef,* the < B P A : B I (the change in

during half the interval between the obfervations by the co-fecant of the meafure of
that half-interval, and turning the products into time, at the rate of 15 degrees to
an hour; the tangent of the latitude, which
is found in the firſt part of the formula,
being left out of the calculation to render
the table general for all latitudes : confequently, the numbers taken out of this
table muſt be multiplied by the tangent of
the geographical latitude; and, if that latitude be ſouth, it will change the ſign of
the equation, becauſe the table is adapted
to northern latitudes.

The ſecond part, which conſiſts of the
continual products formed by multiplying
the change in the ſun's declination during

the ſun's declination) :: cotan. $<$ B: ſin. B P. Confequently,
$$<BPA = \frac{BI \times \text{cotan.} <B}{\text{ſin. } BP}.$$ But it is ſhewn by the writers on
ſpherical trigonometry, that, putting 1 for the radius, the cotan.
of $<B = \dfrac{\text{ſin. } PB \times \text{cot. } PK}{\text{ſin. } <P} \pm \dfrac{\text{coſ. } PB \times \text{coſ. } <P}{\text{ſin. } <P}$. Therefore the
$$<BPA = \frac{BI \times \text{ſin. } PB \times \text{cot. } PK}{\text{ſin. } PB \times \text{ſin. } <P} \pm \frac{BI \times \text{coſ. } PB \times \text{coſ. } <P}{\text{ſin. } PB \times \text{ſin. } <P}.$$
Or, dividing by ſin. P B, and putting the cofecant of $<P$ for
$\dfrac{1}{\text{ſin. } <P}$, in the firſt term, and cotang. for $\dfrac{\text{coſine}}{\text{ſine}}$, in the ſecond
term, we obtain $<BPA = BI \times$ co-fec. $<P \times$ cot. $PK \pm$
$BI \times$ cot. $PB \times$ cot. $<P$: in which, BI is conſidered as
affirmative when PB, the ſun's diſtance from the elevated pole,
is increaſing ; and negative when it is decreaſing.

half the interval between the obfervations, by the co-tangent of the meafure of that half-interval, and by the tangent of the fun's declination, is common to all latitudes; and thefe products, turned into time, are contained in Table II.

Thefe tables may be made to depend either on the fun's longitude or its declination; but as the fun's declination is liable to fome objections, and no advantage can be obtained by ufing it, I have made them depend on the fun's longitude, as ufual.

The method of taking numbers out of the tables will be beft explained by an example:

> Let the fun's longitude be 2ˢ 13° 43′.
> The half-interval 5ʰ 17′ 16″ = 5ʰ 17′,3.
> And the latitude of the place 51° 31′ N.

Firft in Table I. if the fun's longitude be fuppofed 2ˢ 10°, the equation for the half-interval 5ʰ 10′ will be 7″,44; and for the half-interval 5ʰ 20′ it will be 7″,62 ; the difference is 0″,18. Now 10′ is to 7′,3 as 0″,18 is to 0″,13; which being added to 7″,44, the former of them, becaufe the equation increafes, gives 7″,57 for the equation when the fun's longitude is 2ˢ 10°.

If the fun's longitude be taken 2ˢ 15°, the equation will be 5″,65, fuppofing the

half-interval 5ʰ 10′, and 5″,78, ſuppoſing
it 5ʰ 20′: the difference is 0″,13. And 10
is to 7′,3 as 0″,13 is to 0″,09; which being added to the former, gives 5″,74 for
the equation when the ſun's longitude is
2ˢ 15°. The difference between this and
7″,57, the equation when the ſun's longitude was ſuppoſed 2ˢ 10°, is 1″,83. And
300′, (5°) is to 223′, (3° 43′) as 1″,83 is to
1″,36: which being ſubtracted from 7″,57
beeauſe the equation decreaſes this way,
leaves 6″,21 for that part of the equation
which is contained in Table I. But this
number muſt be multiplied by the tangent
of the latitude (51° 31′ N.). I, therefore,
find its logarithm, in Table XVIII. of the
Requiſite Tables, which is 0.79309, and
add to it 0.09965, the tangent of 51° 31′,
in Table XIX. which makes 0.89274, the
logarithm of 7″,81 in Table XVIII, the firſt
part of the equation; which muſt be ſubtracted from the middle time becauſe the
ſign in the table is —. If the latitude had
been ſouth, it would have changed the ſign
to +.

In Table II. ſuppoſing the ſun's longitude to be 2ˢ 10° the equation is 0″,65
when the half-interval is 5ʰ 10′, and 0″,53
when it is 5ʰ 20′; the difference is 0″,12.

And 10' is to 7',3 as 0",12 is to 0",09; which being subtracted from 0",65, because the equation decreases, leaves 0",56 for the equation when the sun's longitude is 2ˢ 10°. Again, the sun's longitude being taken 2ˢ 15°, the equation will be 0",51, when the half-interval is 5ʰ 10', and 0",42 when it is 5ʰ 20'; the difference between them is 0",09. And 10' is to 7',3 as 0",09 is to 0",07; which being subtracted from 0",51, leaves 0",44 for the equation when the sun's longitude is 2ˢ 15°. The difference between 0",44 and 0",56 (the equation when the sun's longitude is 2ˢ 10°) is 0",12. And 300', (5°) is to 223', (3° 43') as 0",12 is to 0",09; which being taken from 0",56, because the equation is decreasing this way also, leaves 0",47 for the proper equation from Table II. which is additive to the middle time, because the sign in the table is +. And the difference between this and 7",81, the equation from Table I. (because they have contrary signs), namely, 7",34 is the equation sought: subtractive, because the greater part is so.

Another example will make every thing relative to these tables perfectly plain to the meanest capacity.

On the 7th of October 1793, the following times, when the fun had equal altitudes were obferved at the Cape of Good Hope, latitude 33° 56′ S. longitude 18° 23′ eaſt.

Ther-mome.	Lower Wire.	Middle Wire.	Upper Wire.		
57° {	21′ 38″,8	21ʰ 22′ 32″,6	23′ 26″,0	☉'s U. L.	} Morn.
	25 31 ,2	21 26 24 ,4	27 17 ,8	☉'s L. L.	
62° {	33 11 ,2	2 32 17 ,6	31 24 ,4	☉'s L. L.	} Aftern.
	37 3 ,2	2 36 9 ,8	35 16 ,6	☉'s U. L.	

	Sun's Upper Limb.			Sun's Lower Limb.			
H. M. S.	H. M. S.	H. M. S.	H. M. S.	H. M. S.	H. M. S.	H. M. S.	
2 37 3,2 24	2 36 9,8 24	2 35 16,6 24	2 33 11,2 24	2 32 17,6 24	2 31 24,4 24		T. of aft obs. Add.
26 37 3,2	26 36 9,8	26 35 16,6	26 33 11,2	26 32 17,6	26 31 24,4		Af.obs. +24h
21 21 38,8	21 22 32.6	21 23 26,0	21 25 31,2	21 26 24,4	21 27 17,8		T.of mor.obs.
5 15 24,4	5 13 37,2	5 11 50,6	5 7 40,0	5 5 53,2	5 4 6,6		Interval.
2 37 42,2	2 36 48,6	2 35 55,3	2 33 50,0	2 32 56,6	2 32 3,3		Half-interv,
23 59 21,0 ,2	23 59 21,2	23 59 21,3	23 59 21,2	23 59 21,0	23 59 21,1		T. n. nearly.
	,3	Sun's long. for noon at Greenwich, October 7th .... 6' 14° 39',0					
	,2 ,0	18° 23′ E. longitude under 2′ 28″ (hourly motion ☉) gives ... — 3,0					
	,1	Sun's longitude for noon at the Cape 6 14 36,0					
6),8							
23 59 21,13	The mean: and 2h 34′ 52″,⅔ is the mean half-interval.						

Firſt ; in Table I. with ☉'s long. 6' 10° and half-interval 2h 30′ the equa. is + 15″,92
☉'s long. 6 10 and half-interval 2h 40′ the equa. is + 16 ,08
The difference is   0 ,16

Now 10′ is to 4′⅛, as 0″,16 is to 0″,08; which being added to 15″,92, becaufe the equation increafes with the interval, gives +16″,00 for the equation when the fun's longitude is 6′ 10°. Again,

☉'s long. 6′ 15° half-int. 2ʰ 30′ the equa. will be +15″,71
☉'s long. 6′ 15° half-int. 2ʰ 40′ the equation is +15 ,87
The difference is 0 ,16

And 10′ is to 4′⅛, as 0″,16 is to 0″,08; which being added to 15″,71, gives +15″,79 for the equation when the fun's longitude is 6′ 15°. The difference between this and 16″,00, the equation when the fun's longitude was 6′ 10°, is 0″,21. Then 300′, (5°) is to 276′, (4° 36′) as 0″,21 is to 0″,19; which being fubtracted from 16,″00, becaufe the equation decreafes this way, gives +15″,81 for the equation from Table I. But this number muft be multiplied by the tangent of the latitude (33° 56′ S.) I therefore find its logarithm, which is 1.19893, and add 9.82790 (the tangent of the latitude) to it; and the fum (1.02683) will be the logarithm of 10″,64, the firft part of the equation; which is fubtractive, becaufe the latitude being fouth changes the fign from + to —.

In Table II. suppofing the fun's longitude to be 6ˢ 10° and the half-interval 2ʰ 30′; the equation will be — 0″,87; and, if the longitude remain the fame, and the half-interval be taken 2ʰ 40′, it will be —0″,85. The difference is 0″,02; and 10′ is to 4′⅓ is 0″,02 is to 0″,01; which being fubtracted from — 0″,87, becaufe this part of the equation decreafes as the half-interval increafes, gives —0″,86 for the equation, when the fun's longitude is 6ˢ 10°. If the fun's longitude be fuppofed 6ˢ 15°, the equation will be — 1″,29, when the half-interval is 2ʰ 30′, and — 1″,26 when it is 2ʰ 40′. The difference between them is 0″,03; and 10′ is to 4⅔ is 0″,03 is to 0″,01; which being fubtracted from —1″,29, gives —1″,28 for the equation when the fun's longitude is 6ˢ 15°. The difference between this and — 0″,86 (the equation when the fun's longitude is 6ˢ 10°) is 0″,42: and 300′, (5°) is to 276′, (4° 36′) as 0″,42 is to 0″,39; which being added to — 0′,86, becaufe the equation increafes this way, gives —1,″25 for the fecond part of the equation. And the two parts being added together, becaufe they are both —, gives —11″,89 for the whole equation. This being fubtracted from 23^h $59'$ $21'',13$, the time of noon nearly, gives 23^h $59'$ $9'',24$ for the true time of apparent noon by the watch.

The equation of time for noon at Green. is —	$12'$	$19'',3$
And 24^h:1^h $14'$(lon. in time):: $16'',7$(d. diff.) : —		0 ,8
Equation of time for noon at the Cape ... —	12 18	,5
Which being fubtracted from	24^h	
Leaves the mean time of apparent noon $=$	23 47 41	,5
Time of noon by the watch	23 59 9	,2
Watch too faft for mean time	11 27	,7

REMARK.

When the clock's rate of going differs very confiderably from mean folar time, and great accuracy is required, the half-interval muft be corrected according to the following proportion, viz. As the time fhewn by the clock in 24 hours is to 24 hours, fo is the half-interval by the clock to the half-interval in mean folar time. And, when the equation is found, it muft be reduced into time of the clock's rate of going, by faying, as 24 hours is to the time fhewn by the clock in 24 hours, fo is the equation found by the preceding directions. to the equation required in this cafe. But unlefs very great accuracy be infifted on, and the clock's rate of going differ more thon 4 or 5 minutes a-day from mean folar time, this correction can never be necef-fary.

TABLES

OF

EQUATIONS

TO

EQUAL ALTITUDES.

TABLE I. *Equations to Equal Altitudes.* 97

⊙'s Long.	Half Interval between the Observations.						
	H. M. I.30	H. M. I.40	H. M. I.50	H. M. II.0	H. M. II.10	H. M. II.20	H. M. II.30
S. D.	S.	S.	S.	S.	S.	S.	S.
O — 0	15,47	15,57	15,68	15,79	15,92	16,06	16,21
5	15,38	15,48	15,58	15,70	15,83	15,97	16,12
10	15,19	15,28	15,38	15,50	15,62	15,76	15,91
15	14,90	14,99	15,09	15,20	15,32	15,46	15,61
20	14,51	14,60	14,69	14,80	14,92	15,05	15,20
25	14,02	14,10	14,20	14,31	14,42	14,55	14,69
I — 0	13,44	13,53	13,62	13,72	13,83	13,95	14,08
5	12,76	12,84	12,93	13,03	13,14	13,25	13,37
10	11,99	12,06	12,14	12,23	12,33	12,44	12,56
15	11,12	11,19	11,27	11,35	11,44	11,54	11,65
20	10,16	10,22	10,29	10,37	10,46	10,55	10,65
25	9,12	9,17	9,23	9,30	9,38	9,46	9,55
II — 0	7,99	8,04	8,09	8,15	8,22	8,29	8,37
5	6,78	6,82	6,87	6,92	6,98	7,04	7,11
10	5,51	5,55	5,59	5,63	5,67	5,72	5,78
15	4,18	4,21	4,24	4,27	4,30	4,34	4,38
20	2,81	2,83	2,85	2,87	2,90	2,92	2,95
25	1,41	1,42	1,43	1,44	1,46	1,47	1,48
III + 0	0,00	0,00	0,00	0,00	0,00	0,00	0,00
5	1,41	1,42	1,43	1,44	1,45	1,47	1,48
10	2,81	2,83	2,85	2,87	2,89	2,92	2,95
15	4,17	4,20	4,23	4,26	4,29	4,33	4,37
20	5,49	5,52	5,56	5,60	5,65	5,70	5,75
25	6,75	6,79	6,84	6,89	6,95	7,01	7,07
IV + 0	7,94	7,99	8,05	8,11	8,17	8,24	8,32
5	9,06	9,11	9,17	9,24	9,32	9,40	9,49
10	10,09	10,15	10,22	10,30	10,38	10,47	10,57
15	11,03	11,10	11,18	11,26	11,35	11,45	11,56
20	11,89	11,96	12,04	12,13	12,23	12,33	12,45
25	12,65	12,72	12,81	12,91	13,01	13,12	13,25
V + 0	13,31	13,40	13,49	13,59	13,70	13,82	13,95
5	13,88	13,97	14,07	14,17	14,28	14,41	14,54
10	14,35	14,44	14,54	14,65	14,77	14,90	15,04
15	14,74	14,83	14,93	15,04	15,16	15,30	15,44
20	15,03	15,12	15,22	15,33	15,46	15,59	15,74
25	15,22	15,31	15,41	15,53	15,65	15,79	15,94
VI + 0	15,31	15,40	15,51	15,62	15,75	15,89	16,04

		Half Interval between the Observations.						
⊙'s Long.		H. M. II.40	H. M. II.50	H. M. III.0	H. M. III.10	H. M. III.20	H. M. III.30	H. M. III.40
S.	D.	S.	S.	S.	S.	S.	S.	S.
O	− 0	16,38	16,56	16,75	16,96	17,18	17,42	17,67
	5	16,28	16,46	16,65	16,86	17,08	17,32	17,57
	10	16,08	16,25	16,44	16,64	16,86	17,10	17,35
	15	15,77	15,94	16,12	16,32	16,54	16,77	17,01
	20	15,35	15,52	15,70	15,89	16,10	16,32	16,56
	25	14,84	15,00	15,17	15,36	15,56	15,78	16,01
I	− 0	14,22	14,37	14,54	14,72	14,92	15,13	15,35
	5	13,51	13,65	13,81	13,98	14,17	14,36	14,57
	10	12,69	12,83	12,98	13,14	13,31	13,49	13,69
	15	11,77	11,90	12,04	12,19	12,35	12,52	12,70
	20	10,75	10,87	11,00	11,14	11,28	11,44	11,61
	25	9,65	9,76	9,87	9,99	10,12	10,26	10,41
II	− 0	8,46	8,55	8,65	8,76	8,87	8,99	9,12
	5	7,18	7,26	7,35	7,44	7,54	7,64	7,75
	10	5,84	5,90	5,97	6,04	6,12	6,21	6,30
	15	4,43	4,48	4,53	4,59	4,65	4,71	4,78
	20	2,98	3,01	3,05	3,09	3,13	3,17	3,22
	25	1,50	1,51	1,53	1,55	1,57	1,59	1,62
III	+ 0	0,00	0,00	0,00	0,00	0,00	0,00	0,00
	5	1,49	1,51	1,53	1,55	1,57	1,59	1,61
	10	2,98	3,01	3,04	3,08	3,12	3,16	3,21
	15	4,42	4,47	4,52	4,57	4,63	4,70	4,77
	20	5,81	5,87	5,94	6,02	6,10	6,18	6,27
	25	7,14	7,22	7,31	7,40	7,50	7,60	7,71
IV	+ 0	8,41	8,50	8,60	8,71	8,82	8,94	9,07
	5	9,59	9,69	9,80	9,93	10,06	10,20	10,34
	10	10,68	10,80	10,93	11,06	11,20	11,36	11,53
	15	11,68	11,81	11,94	12,09	12,25	12,42	12,60
	20	12,58	12,72	12,86	13,02	13,19	13,38	13,57
	25	13,38	13,53	13,69	13,86	14,04	14,23	14,44
V	+ 0	14,09	14,24	14,41	14,59	14,78	14,98	15,20
	5	14,69	14,85	15,03	15,21	15,41	15,62	15,85
	10	15,20	15,36	15,54	15,73	15,94	16,16	16,39
	15	15,60	15,77	15,95	16,15	16,36	16,59	16,83
	20	15,90	16,07	16,26	16,46	16,68	16,91	17,16
	25	16,10	16,28	16,47	16,67	16,89	17,12	17,37
VI	+ 0	16,20	16,38	16,57	16,78	17,00	17,23	17,48

TABLE I. *Equations to Equal Altitudes.* 99

⊙'s Long.		Half Interval between the Observations.						
		H. M. III.50	H. M. IV. 0	H. M. IV.10	H. M. IV.20	H. M. IV.30	H. M. IV.40	H. M. IV.50
S.	D.	S.	S.	S.	S.	S.	S.	S.
O	− 0	17,94	18,23	18,54	18,87	19,22	19,60	20,01
	5	17,84	18,13	18,44	18,77	19,12	19,49	19,89
	10	17,62	17,90	18,20	18,53	18,88	19,25	19,64
	15	17,27	17,55	17,85	18,17	18,51	18,87	19,26
	20	16,82	17,09	17,38	17,69	18,02	18,37	18,75
	25	16,26	16,52	16,80	17,10	17,42	17,76	18,12
I	− 0	15,59	15,84	16,11	16,40	16,70	17,03	17,38
	5	14,80	15,04	15,29	15,56	15,86	16,17	16,50
	10	13,90	14,13	14,37	14,63	14,90	15,19	15,50
	15	12,90	13,11	13,33	13,57	13,82	14,09	14,38
	20	11,79	11,98	12,18	12,40	12,63	12,88	13,14
	25	10,57	10,74	10,93	11,12	11,33	11,55	11,79
II	− 0	9,26	9,41	9,57	9,74	9,93	10,12	10,33
	5	7,87	8,00	8,14	8,28	8,43	8,60	8,77
	10	6,40	6,50	6,61	6,72	6,85	6,99	7,13
	15	4,85	4,93	5,01	5,10	5,20	5,30	5,41
	20	3,27	3,32	3,37	3,43	3,50	3,57	3,64
	25	1,64	1,67	1,70	1,73	1,76	1,79	1,83
III	+ 0	0,00	0,00	0,00	0,00	0,00	0,00	0,00
	5	1,64	1,66	1,69	1,72	1,76	1,79	1,83
	10	3,26	3,31	3,37	3,43	3,50	3,56	3,63
	15	4,84	4,92	5,00	5,09	5,19	5,29	5,40
	20	6,37	6,47	6,58	6,70	6,82	6,96	7,10
	25	7,83	7,96	8,10	8,24	8,39	8,56	8,73
IV	+ 0	9,21	9,36	9,52	9,69	9,87	10,06	10,27
	5	10,50	10,68	10,86	11,05	11,26	11,48	11,71
	10	11,70	11,89	12,10	12,31	12,54	12,79	13,05
	15	12,79	13,00	13,22	13,46	13,71	13,98	14,27
	20	13,78	14,01	14,25	14,50	14,77	15,06	15,37
	25	14,66	14,90	15,15	15,42	15,71	16,02	16,35
V	+ 0	15,44	15,69	15,96	16,24	16,54	16,86	17,21
	5	16,10	16,36	16,64	16,94	17,26	17,59	17,95
	10	16,64	16,91	17,20	17,51	17,84	18,19	18,56
	15	17,09	17,37	17,67	17,98	18,31	18,67	19,06
	20	17,42	17,70	18,00	18,32	18,67	19,04	19,43
	25	17,64	17,93	18,23	18,56	18,91	19,28	19,67
VI	+ 0	17,75	18,04	18,35	18,68	19,03	19,40	19,79

| ⊙'s Long. | | Half Interval between the Observations. | | | | | | |
		H. M. V. 0	H. M. V. 10	H. M. V. 20	H. M. V. 30	H. M. V. 40	H. M. V. 50	H. M. VI. 0
S.	D.	S.	S.	S.	S.	S.	S.	S.
O	− 0	20,44	20,90	21,38	21,90	22,46	23,05	23,69
	5	20,32	20,77	21,25	21,77	22,32	22,91	23,55
	10	20,06	20,51	20,99	21,50	22,04	22,62	23,25
	15	19,67	20,11	20,58	21,08	21,62	22,19	22,80
	20	19,16	19,59	20,05	20,54	21,06	21,61	22,21
	25	18,51	18,93	19,37	19,84	20,34	20,88	21,46
I	− 0	17,75	18,15	18,57	19,02	19,50	20,03	20,57
	5	16,85	17,23	17,63	18,06	18,52	19,01	19,53
	10	15,83	16,19	16,57	16,97	17,40	17,86	18,35
	15	14,69	15,02	15,37	15,74	16,14	16,57	17,03
	20	13,42	13,72	14,04	14,38	14,75	15,14	15,56
	25	12,04	12,31	12,59	12,90	13,23	13,58	13,96
II	− 0	10,55	10,78	11,03	11,30	11,59	11,90	12,23
	5	8,96	9,16	9,37	9,60	9,85	10,11	10,39
	10	7,28	7,44	7,62	7,80	8,00	8,21	8,44
	15	5,53	5,65	5,78	5,92	6,07	6,23	6,41
	20	3,72	3,80	3,89	3,98	4,08	4,19	4,31
	25	1,87	1,91	1,95	2,00	2,05	2,10	2,16
III	+ 0	0,00	0,00	0,00	0,00	0,00	0,00	0,00
	5	1,87	1,91	1,95	2,00	2,05	2,10	2,16
	10	3,71	3,79	3,88	3,97	4,07	4,18	4,30
	15	5,51	5,63	5,77	5,91	6,06	6,22	6,39
	20	7,25	7,41	7,59	7,77	7,97	8,18	8,41
	25	8,92	9,12	9,33	9,56	9,80	10,06	10,34
IV	+ 0	10,49	10,73	10,98	11,25	11,54	11,84	12,16
	5	11,96	12,23	12,52	12,82	13,14	13,49	13,86
	10	13,33	13,63	13,94	14,28	14,65	15,04	15,45
	15	14,57	14,90	15,25	15,62	16,01	16,43	16,89
	20	15,70	16,05	16,42	16,82	17,25	17,71	18,19
	25	16,70	17,07	17,47	17,90	18,35	18,84	19,36
V	+ 0	17,58	17,98	18,40	18,85	19,32	19,83	20,38
	5	18,34	18,75	19,18	19,64	20,14	20,67	21,25
	10	18,96	19,39	19,84	20,32	20,84	21,39	21,97
	15	19,47	19,90	20,36	20,86	21,39	21,96	22,56
	20	19,84	20,28	20,76	21,26	21,80	22,38	23,00
	25	20,09	20,54	21,02	21,53	22,08	22,66	23,29
VI	+ 0	20,22	20,67	21,15	21,67	22,22	22,81	23,44

TABLE I. *Equations to Equal Altitudes.* 101

⊙'s Long.		Half Interval between the Observations.						
		H. M. I. 30	H. M. I. 40	H. M. I. 50	H. M. II. 0	H. M. II. 10	H. M. II. 20	H. M. II. 30
S.	D.	S.	S.	S.	S.	S.	S.	S.
VI +	0	15,31	15,41	15,51	15,62	15,75	15,89	16,04
	5	15,30	15,40	15,50	15,61	15,74	15,88	16,03
	10	15,20	15,29	15,39	15,51	15,63	15,77	15,92
	15	14,99	15,08	15,19	15,30	15,42	15,56	15,71
	20	14,69	14,78	14,88	14,99	15,11	15,24	15,39
	25	14,28	14,37	14,46	14,57	14,69	14,82	14,96
VII +	0	13,76	13,85	13,95	14,05	14,16	14,29	14,42
	5	13,14	13,22	13,31	13,41	13,52	13,64	13,77
	10	12,41	12,49	12,57	12,66	12,76	12,87	13,00
	15	11,57	11,64	11,71	11,80	11,90	12,01	12,12
	20	10,62	10,68	10,75	10,83	10,92	11,02	11,13
	25	9,57	9,63	9,69	9,76	9,84	9,93	10,03
VIII +	0	8,42	8,47	8,53	8,59	8,66	8,74	8,82
	5	7,18	7,22	7,27	7,32	7,38	7,45	7,52
	10	5,85	5,88	5,92	5,97	6,02	6,07	6,13
	15	4,45	4,48	4,51	4,54	4,58	4,62	4,66
	20	3,00	3,02	3,04	3,06	3,09	3,11	3,14
	25	1,51	1,52	1,53	1,54	1,55	1,56	1,58
IX −	0	0,00	0,00	0,00	0,00	0,00	0,00	0,00
	5	1,51	1,52	1,53	1,54	1,55	1,57	1,58
	10	3,00	3,02	3,04	3,07	3,09	3,12	3,15
	15	4,46	4,49	4,52	4,55	4,59	4,63	4,67
	20	5,87	5,90	5,94	5,99	6,04	6,09	6,15
	25	7,20	7,24	7,29	7,35	7,41	7,48	7,55
X −	0	8,46	8,51	8,57	8,63	8,70	8,78	8,86
	5	9,63	9,69	9,75	9,82	9,90	9,99	10,09
	10	10,70	10,76	10,83	10,91	11,00	11,10	11,21
	15	11,66	11,73	11,81	11,90	12,00	12,10	12,21
	20	12,51	12,59	12,68	12,77	12,87	12,99	13,11
	25	13,26	13,34	13,43	13,53	13,64	13,76	13,89
XI −	0	13,89	13,98	14,08	14,18	14,29	14,42	14,55
	5	14,42	14,51	14,61	14,72	14,84	14,97	15,11
	10	14,84	14,93	15,03	15,14	15,27	15,40	15,55
	15	15,15	15,25	15,35	15,46	15,59	15,73	15,88
	20	15,36	15,46	15,57	15,68	15,80	15,94	16,10
	25	15,47	15,56	15,67	15,78	15,91	16,05	16,21
XII −	0	15,48	15,57	15,68	15,79	15,92	16,06	16,22

⊙'s Long.		Half Interval between the Observations.						
		H. M. II. 40	H. M. II. 50	H. M. III. 0	H. M. III.10	H. M. III.20	H. M. III.30	H. M. III.40
S.	D.	S.	S.	S.	S.	S.	S.	S.
VI +	0	16,21	16,38	16,57	16,78	17,00	17,23	17,48
	5	16,20	16,37	16,56	16,77	16,99	17,22	17,47
	10	16,08	16,26	16,45	16,65	16,37	17,10	17,35
	15	15,87	16,04	16,23	16,43	16,64	16,87	17,12
	20	15,55	15,72	15,90	16,09	16,30	16,53	16,77
	25	15,11	15,28	15,46	15,65	15,86	16,08	16,31
VII +	0	14,57	14,73	14,90	15,08	15,28	15,49	15,72
	5	13,91	14,06	14,22	14,40	14,59	14,79	15,01
	10	13,13	13,27	13,42	13,59	13,77	13,96	14,17
	15	12,24	12,37	12,51	12,67	12,84	13,02	13,21
	20	11,24	11,36	11,49	11,63	11,79	11,96	12,13
	25	10,13	10,24	10,36	10,48	10,62	10,77	10,93
VIII +	0	8,91	9,01	9,11	9,22	9,34	9,47	9,61
	5	7,59	7,67	7,76	7,86	7,96	8,07	8,19
	10	6,19	6,26	6,33	6,41	6,49	6,58	6,68
	15	4,71	4,76	4,82	4,88	4,94	5,01	5,08
	20	3,18	3,21	3,25	3,29	3,33	3,38	3,43
	25	1,60	1,62	1,64	1,66	1,68	1,70	1,73
IX −	0	0,00	0,00	0,00	0,00	0,00	0,00	0,00
	5	1,60	1,62	1,64	1,66	1,68	1,70	1,73
	10	3,18	3,21	3,25	3,28	3,33	3,38	3,43
	15	4,72	4,77	4,83	4,89	4,95	5,02	5,09
	20	6,21	6,28	6,35	6,43	6,51	6,60	6,70
	25	7,62	7,70	7,79	7,89	8,00	8,11	8,23
X −	0	8,95	9,05	9,15	9,27	9,39	9,52	9,66
	5	10,19	10,30	10,42	10,55	10,69	10,83	10,99
	10	11,32	11,44	11,58	11,72	11,87	12,04	12,21
	15	12,34	12,48	12,62	12,77	12,94	13,12	13,31
	20	13,24	13,39	13,54	13,71	13,89	14,08	14,29
	25	14,03	14,18	14,35	14,53	14,72	14,92	15,14
XI −	0	14,70	14,86	15,04	15,23	15,42	15,63	15,86
	5	15,26	15,43	15,61	15,80	16,01	16,23	16,47
	10	15,71	15,88	16,06	16,26	16,48	16,71	16,95
	15	16,04	16,21	16,40	16,60	16,82	17,06	17,31
	20	16,26	16,44	16,63	16,83	17,05	17,29	17,54
	25	16,37	16,55	16,74	16,95	17,17	17,41	17,66
XII −	0	16,38	16,56	16,75	16,96	17,18	17,42	17,67

TABLE I. *Equations to Equal Altitudes.* 103

⊙'s Long.		Half Interval between the Observations.						
		H. M. III.50	H. M. IV. 0	H. M. IV.10	H. M. IV.20	H. M. IV.30	H. M. IV.40	H. M. IV.50
S.	D.	S.	S.	S.	S.	S.	S.	S.
VI	+ 0	17,75	18,04	18,35	18,67	19,02	19,40	19,80
	5	17,74	18,03	18,34	18,66	19,01	19,39	19,79
	10	17,62	17,91	18,22	18,54	18,88	19,25	19,65
	15	17,39	17,67	17,97	18,29	18,63	18,99	19,38
	20	17,03	17,31	17,60	17,92	18,25	18,61	18,99
	25	16,56	16,82	17,11	17,42	17,75	18,10	18,47
VII	+ 0	15,96	16,22	16,49	16,79	17,11	17,44	17,80
	5	15,24	15,49	15,75	16,03	16,33	16,65	16,99
	10	14,39	14,62	14,87	15,14	15,42	15,72	16,04
	15	13,41	13,63	13,86	14,11	14,37	14,65	14,95
	20	12,31	12,51	12,72	12,95	13,19	13,45	13,73
	25	11,09	11,27	11,46	11,67	11,89	12,12	12,37
VIII	+ 0	9,76	9,92	10,09	10,27	10,46	10,66	10,88
	5	8,32	8,45	8,59	8,75	8,92	9,09	9,27
	10	6,78	6,89	7,01	7,14	7,27	7,41	7,56
	15	5,16	5,24	5,33	5,43	5,53	5,64	5,76
	20	3,48	3,54	3,60	3,66	3,73	3,80	3,88
	25	1,75	1,78	1,81	1,84	1,88	1,91	1,95
IX	− 0	0,00	0,00	0,00	0,00	0,00	0,00	0,00
	5	1,75	1,78	1,81	1,84	1,88	1,92	1,95
	10	3,48	3,54	3,60	3,66	3,73	3,81	3,89
	15	5,17	5,26	5,35	5,44	5,54	5,65	5,77
	20	6,80	6,91	7,03	7,16	7,29	7,43	7,59
	25	8,36	8,49	8,63	8,79	8,95	9,12	9,31
X	− 0	9,81	9,97	10,14	10,32	10,51	10,72	10,94
	5	11,16	11,34	11,54	11,74	11,96	12,20	12,45
	10	12,40	12,60	12,82	13,05	13,29	13,55	13,83
	15	13,52	13,74	13,97	14,22	14,49	14,77	15,07
	20	14,51	14,74	14,99	15,26	15,55	15,86	16,18
	25	15,37	15,62	15,89	16,17	16,47	16,79	17,14
XI	− 0	16,11	16,37	16,65	16,95	17,26	17,60	17,96
	5	16,72	16,99	17,28	17,59	17,92	18,27	18,65
	10	17,21	17,49	17,79	18,10	18,44	18,80	19,19
	15	17,57	17,86	18,16	18,48	18,83	19,20	19,59
	20	17,81	18,10	18,41	18,74	19,09	19,46	19,86
	25	17,94	18,23	18,54	18,87	19,22	19,60	20,00
XII	− 0	17,95	18,24	18,55	18,88	19,23	19,61	20,01

⊙'s Long.		Half Interval between the Observations.						
		H. M. V. 0	H. M. V. 10	H. M. V. 20	H. M. V. 30	H. M. V. 40	H. M. V. 50	H. M. VI. 0
S.	D.	S.	S.	S.	S.	S.	S.	S.
VI +	0	20,22	20,67	21,15	21,67	22,22	22,81	23,43
	5	20,21	20,66	21,14	21,66	22,21	22,80	23,42
	10	20,07	20,52	21,00	21,51	22,05	22,64	23,26
	15	19,80	20,24	20,71	21,22	21,76	22,33	22,95
	20	19,40	19,83	20,29	20,79	21,32	21,88	22,48
	25	18,86	19,28	19,73	20,21	20,72	21,27	21,85
VII +	0	18,18	18,58	19,02	19,48	19,97	20,50	21,06
	5	17,35	17,74	18,15	18,60	19,07	19,57	20,11
	10	16,38	16,75	17,14	17,56	18,00	18,48	18,99
	15	15,27	15,61	15,98	16,37	16,78	17,23	17,70
	20	14,03	14,34	14,67	15,03	15,41	15,82	16,25
	25	12,64	12,92	13,22	13,54	13,88	14,25	14,65
VIII +	0	11,11	11,36	11,63	11,91	12,21	12,54	12,89
	5	9,47	9,68	9,91	10,15	10,41	10,68	10,98
	10	7,72	7,89	8,08	8,28	8,49	8,71	8,95
	15	5,88	6,01	6,15	6,30	6,46	6,63	6,81
	20	3,96	4,05	4,14	4,24	4,35	4,47	4,59
	25	1,99	2,04	2,09	2,14	2,19	2,25	2,31
IX −	0	0,00	0,00	0,00	0,00	0,00	0,00	0,00
	5	1,99	2,04	2,09	2,14	2,19	2,25	2,31
	10	3,97	4,06	4,15	4,25	4,36	4,48	4,60
	15	5,89	6,02	6,17	6,32	6,48	6,65	6,83
	20	7,75	7,92	8,11	8,31	8,52	8,74	8,98
	25	9,51	9,72	9,95	10,19	10,45	10,73	11,02
X −	0	11,17	11,42	11,69	11,97	12,28	12,60	12,94
	5	12,72	13,00	13,30	13,62	13,97	14,34	14,73
	10	14,12	14,44	14,78	15,14	15,52	15,93	16,37
	15	15,39	15,74	16,11	16,50	16,92	17,37	17,84
	20	16,52	16,89	17,29	17,71	18,16	18,64	19,15
	25	17,51	17,90	18,32	18,76	19,24	19,75	20,29
XI −	0	18,35	18,76	19,19	19,66	20,16	20,69	21,26
	5	19,05	19,47	19,92	20,41	20,93	21,48	22,07
	10	19,60	20,04	20,50	21,00	21,54	22,11	22,72
	15	20,01	20,45	20,94	21,45	21,99	22,57	23,20
	20	20,28	20,73	21,22	21,74	22,29	22,88	23,51
	25	20,43	20,89	21,37	21,89	22,45	23,04	23,68
XII −	0	20,44	20,90	21,38	21,90	22,46	23,05	23,69

TABLE II. *Equations to Equal Altitudes.* 105

⊙'s Long.		Half Interval between the Observations.						
		H. M. I . 30	H. M. I. 40	H. M. I. 50	H. M. II. 0	H. M. II. 10	H. M. II. 20	H. M. II. 30
S.	D.	S.	S.	S.	S.	S.	S.	S.
O +	0	0,00	0,00	0,00	0,00	0,00	0,00	0,00
	5	0,49	0,49	0,48	0,47	0,46	0,45	0,44
	10	0,97	0,96	0,95	0,93	0,91	0,90	0,88
	15	1,43	1,41	1,39	1,36	1,34	1,31	1,28
	20	1,84	1,82	1,79	1,76	1,73	1,70	1,66
	25	2,21	2,18	2,15	2,12	2,08	2,04	1,99
I +	0	2,52	2,49	2,45	2,41	2,37	2,32	2,27
	5	2,77	2,73	2,69	2,65	2,60	2,55	2,49
	10	2,93	2,89	2,85	2,81	2,76	2,70	2,64
	15	3,01	2,97	2,93	2,88	2,83	2,77	2,71
	20	3,01	2,97	2,93	2,88	2,83	2,77	2,71
	25	2,91	2,87	2,83	2,78	2,73	2,67	2,61
II +	0	2,71	2,68	2,64	2,60	2,55	2,50	2,44
	5	2,42	2,39	2,36	2,32	2,28	2,23	2,18
	10	2,06	2,03	2,00	1,97	1,93	1,89	1,85
	15	1,61	1,59	1,57	1,54	1,51	1,48	1,45
	20	1,11	1,10	1,08	1,06	1,04	1,02	1,00
	25	0,57	0,56	0,55	0,54	0,53	0,52	0,51
III —	0	0,00	0,00	0,00	0,00	0,00	0,00	0,00
	5	0,56	0,56	0,55	0,54	0,53	0,52	0,51
	10	1,11	1,10	1,08	1,06	1,04	1,02	1,00
	15	1,61	1,59	1,57	1,54	1,51	1,48	1,45
	20	2,05	2,02	1,99	1,96	1,92	1,88	1,84
	25	2,41	2,38	2,35	2,31	2,27	2,22	2,17
IV —	0	2,70	2,66	2,62	2,58	2,53	2,48	2,43
		2,89	2,85	2,81	2,76	2,71	2,66	2,60
	10	2,99	2,95	2,90	2,85	2,80	2,75	2,69
	15	2,99	2,95	2,91	2,86	2,81	2,75	2,69
	20	2,91	2,87	2,83	2,78	2,73	2,68	2,62
	25	2,74	2,70	2,66	2,62	2,57	2,52	2,47
V —	0	2,50	2,47	2,43	2,39	2,35	2,30	2,25
	5	2,19	2,16	2,13	2,09	2,05	2,01	1,97
	10	1,82	1,80	1,77	1,74	1,71	1,68	1,64
	15	1,41	1,39	1,37	1,35	1,33	1,30	1,27
	20	0,96	0,95	0,94	0,92	0,90	0,88	0,86
	25	0,49	0,48	0,47	0,47	0,46	0,45	0,44
VI —	0	0,00	0,00	0,00	0,00	0,00	0,00	0,00

P

⊙'s Long.		Half Interval between the Observations.						
		H. M. II. 40	H. M. II. 50	H. M. III. 0	H. M. III.10	H. M. III.20	H. M. III.30	H. M. III.40
S.	D.	S.	S.	S.	S.	S.	S.	S.
O	+ 0	0,00	0,00	0,00	0,00	0,00	0,00	0,00
	5	0,43	0,42	0,41	0,40	0,38	0,37	0,35
	10	0,85	0,83	0,81	0,78	0,75	0,72	0,69
	15	1,25	1,22	1,18	1,14	1,10	1,06	1,01
	20	1,62	1,58	1,53	1,48	1,43	1,37	1,31
	25	1,94	1,89	1,83	1,77	1,71	1,64	1,57
I	+ 0	2,21	2,15	2,09	2,02	1,95	1,87	1,79
	5	2,43	2,36	2,29	2,22	2,14	2,05	1,96
	10	2,57	2,50	2,43	2,35	2,26	2,17	2,08
	15	2,64	2,57	2,50	2,42	2,33	2,24	2,14
	20	2,64	2,57	2,49	2,41	2,32	2,23	2,13
	25	2,55	2,48	2,41	2,33	2,24	2,15	2,06
II	+ 0	2,38	2,31	2,24	2,17	2,09	2,01	1,92
	5	2,13	2,07	2,01	1,94	1,87	1,80	1,72
	10	1,80	1,75	1,70	1,65	1,59	1,53	1,46
	15	1,41	1,37	1,33	1,29	1,24	1,19	1,14
	20	0,98	0,95	0,92	0,89	0,86	0,83	0,79
	25	0,50	0,48	0,47	0,45	0,44	0,42	0,40
III	− 0	0,00	0,00	0,00	0,00	0,00	0,00	0,00
	5	0,49	0,48	0,47	0,45	0,44	0,42	0,40
	10	0,97	0,94	0,91	0,88	0,85	0,82	0,78
	15	1,41	1,37	1,33	1,29	1,24	1,19	1,14
	20	1,80	1,75	1,70	1,64	1,58	1,52	1,45
	25	2,12	2,06	2,00	1,93	1,86	1,79	1,71
IV	− 0	2,37	2,30	2,23	2,16	2,08	2,00	1,91
	5	2,54	2,47	2,39	2,31	2,23	2,14	2,05
	10	2,62	2,55	2,47	2,39	2,31	2,22	2,12
	15	2,62	2,55	2,48	2,40	2,31	2,22	2,12
	20	2,55	2,48	2,41	2,33	2,25	2,16	2,06
	25	2,41	2,34	2,27	2,20	2,12	2,03	1,94
V	− 0	2,19	2,13	2,07	2,00	1,93	1,85	1,77
	5	1,92	1,87	1,81	1,75	1,69	1,62	1,55
	10	1,60	1,56	1,51	1,46	1,41	1,35	1,29
	15	1,24	1,21	1,17	1,13	1,09	1,05	1,00
	20	0,84	0,82	0,80	0,77	0,74	0,71	0,68
	25	0,43	0,42	0,40	0,39	0,38	0,36	0,35
VI	− 0	0,00	0,00	0,00	0,00	0,00	0,00	0,00

TABLE II. *Equations to Equal Altitudes.* 107

⊙'s Long.		Half Interval between the Observations.						
		H. M. III.50	H. M. IV. 0	H. M. IV.10	H. M. IV.20	H. M. IV.30	H. M. IV.40	H. M. IV.50
S.	D.	S.	S.	S.	S.	S.	S.	S.
O	+ 0	0,00	0,00	0,00	0,00	0,00	0,00	0,00
	5	0,33	0,31	0,30	0,28	0,25	0,23	0,21
	10	0,66	0,62	0,58	0,54	0,50	0,46	0,41
	15	0,96	0,91	0,85	0,79	0,73	0,67	0,60
	20	1,24	1,17	1,10	1,03	0,95	0,87	0,78
	25	1,49	1,41	1,32	1,23	1,14	1,04	0,93
I	+ 0	1,70	1,61	1,51	1,41	1,30	1,18	1,06
	5	1,86	1,76	1,65	1,54	1,42	1,29	1,16
	10	1,98	1,87	1,76	1,64	1,51	1,37	1,23
	15	2,03	1,92	1,80	1,68	1,55	1,41	1,27
	20	2,03	1,92	1,80	1,68	1,55	1,41	1,26
	25	1,96	1,85	1,74	1,62	1,49	1,36	1,22
II	+ 0	1,83	1,73	1,62	1,51	1,39	1,27	1,14
	5	1,64	1,55	1,45	1,35	1,25	1,14	1,02
	10	1,39	1,31	1,23	1,15	1,06	0,96	0,86
	15	1,09	1,03	0,97	0,90	0,83	0,76	0,68
	20	0,75	0,71	0,67	0,62	0,57	0,52	0,47
	25	0,38	0,36	0,34	0,32	0,29	0,26	0,24
III	− 0	0,00	0,00	0,00	0,00	0,00	0,00	0,00
	5	0,38	0,36	0,34	0,31	0,29	0,26	0,24
	10	0,75	0,71	0,67	0,62	0,57	0,52	0,47
	15	1,08	1,02	0,96	0,90	0,83	0,76	0,68
	20	1,38	1,31	1,23	1,14	1,05	0,96	0,86
	25	1,63	1,54	1,45	1,35	1,24	1,13	1,02
IV	− 0	1,82	1,72	1,61	1,50	1,38	1,26	1,13
	5	1,95	1,84	1,73	1,61	1,48	1,35	1,21
	10	2,01	1,90	1,79	1,67	1,54	1,40	1,26
	15	2,02	1,91	1,79	1,67	1,54	1,40	1,26
	20	1,96	1,85	1,74	1,62	1,49	1,36	1,22
	25	1,85	1,75	1,64	1,53	1,41	1,28	1,15
V	− 0	1,68	1,59	1,50	1,40	1,29	1,17	1,05
	5	1,48	1,40	1,31	1,22	1,13	1,03	0,92
	10	1,23	1,16	1,09	1,02	0,94	0,86	0,77
	15	0,95	0,90	0,85	0,79	0,73	0,66	0,59
	20	0,65	0,62	0,58	0,54	0,50	0,45	0,40
	25	0,33	0,31	0,29	0,27	0,25	0,23	0,21
VI	− 0	0,00	0,00	0,00	0,00	0,00	0,00	0,00

⊙'s Long.		Half Interval between the Observations.						
		H. M. V. 0	H. M. V. 10	H. M. V. 20	H. M. V. 30	H. M. V. 40	H. M. V. 50	H. M. VI. 0
S.	D.	S.	S.	S.	S.	S.	S.	S.
O	+ 0	0,00	0,00	0,00	0,00	0,00	0,00	0,00
	5	0,18	0,16	0,13	0,10	0,07	0,04	0,00
	10	0,36	0,31	0,25	0,19	0,13	0,07	0,00
	15	0,53	0,45	0,37	0,29	0,20	0,10	0,00
	20	0,68	0,58	0,48	0,37	0,25	0,13	0,00
	25	0,82	0,70	0,57	0,44	0,30	0,16	0,00
I	+ 0	0,93	0,80	0,66	0,51	0,35	0,18	0,00
	5	1,02	0,87	0,71	0,55	0,38	0,19	0,00
	10	1,08	0,92	0,76	0,59	0,41	0,21	0,00
	15	1,12	0,95	0,78	0,60	0,41	0,21	0,00
	20	1,11	0,95	0,78	0,60	0,41	0,21	0,00
	25	1,07	0,91	0,75	0,58	0,40	0,20	0,00
II	+ 0	1,00	0,85	0,70	0,54	0,37	0,19	0,00
	5	0,90	0,77	0,63	0,48	0,33	0,17	0,00
	10	0,76	0,65	0,53	0,41	0,28	0,14	0,00
	15	0,60	0,51	0,42	0,32	0,22	0,11	0,00
	20	0,41	0,35	0,29	0,22	0,15	0,08	0,00
	25	0,21	0,18	0,15	0,11	0,08	0,04	0,00
III	− 0	0,00	0,00	0,00	0,00	0,00	0,00	0,00
	5	0,21	0,18	0,15	0,11	0,08	0,04	0,00
	10	0,41	0,35	0,29	0,22	0,15	0,08	0,00
	15	0,60	0,51	0,42	0,32	0,22	0,11	0,00
	20	0,76	0,65	0,53	0,41	0,28	0,14	0,00
	25	0,90	0,77	0,63	0,48	0,33	0,17	0,00
IV	− 0	0,99	0,85	0,70	0,54	0,37	0,19	0,00
	5	1,06	0,91	0,75	0,58	0,40	0,20	0,00
	10	1,11	0,95	0,78	0,60	0,41	0,21	0,00
	15	1,11	0,95	0,78	0,60	0,41	0,21	0,00
	20	1,10	0,92	0,76	0,58	0,39	0,20	0,00
	25	1,01	0,86	0,71	0,55	0,38	0,19	0,00
V	− 0	0,92	0,79	0,65	0,50	0,34	0,17	0,00
	5	0,81	0,69	0,57	0,44	0,30	0,15	0,00
	10	0,68	0,58	0,47	0,36	0,25	0,13	0,00
	15	0,52	0,45	0,37	0,28	0,19	0,10	0,00
	20	0,35	0,30	0,25	0,19	0,13	0,07	0,00
	25	0,18	0,15	0,13	0,10	0,07	0,04	0,00
VI	− 0	0,00	0,00	0,00	0,00	0,00	0,00	0,00

TABLE II. *Equations to Equal Altitudes.* 109

☉'s Long.		Half Interval between the Observations.						
		H. M. I. 30	H. M. I. 40	H. M. I. 50	H. M. II. 0	H. M. II. 10	H. M. II. 20	H. M. II. 30
s.	D.	S.	S.	S.	S.	S.	S.	S.
VI +	0	0,00	0,00	0,00	0,00	0,00	0,00	0,00
	5	0,49	0,49	0,48	0,47	0,46	0,45	0,44
	10	0,97	0,96	0,95	0,93	0,91	0,89	0,87
	15	1,43	1,42	1,40	1,37	1,35	1,32	1,29
	20	1,86	1,84	1,81	1,78	1,75	1,72	1,68
	25	2,25	2,22	2,19	2,15	2,11	2,07	2,03
VII +	0	2,58	2,55	2,51	2,47	2,43	2,38	2,33
	5	2,85	2,81	2,77	2,72	2,67	2,62	2,56
	10	3,04	3,00	2,95	2,90	2,85	2,79	2,73
	15	3,14	3,10	3,05	3,00	2,94	2,88	2,82
	20	3,14	3,10	3,06	3,01	2,95	2,89	2,83
	25	3,05	3,01	2,97	2,92	2,87	2,81	2,75
VIII +	0	2,86	2,82	2,78	2,73	2,68	2,63	2,57
	5	2,57	2,53	2,49	2,45	2,41	2,36	2,31
	10	2,18	2,15	2,12	2,08	2,05	2,01	1,96
	15	1,71	1,69	1,67	1,64	1,61	1,58	1,54
	20	1,18	1,17	1,15	1,13	1,11	1,09	1,06
	25	0,60	0,60	0,59	0,58	0,57	0,55	0,54
IX −	0	0,00	0,00	0,00	0,00	0,00	0,00	0,00
	5	0,60	0,60	0,59	0,58	0,57	0,55	0,54
	10	1,18	1,17	1,15	1,13	1,11	1,09	1,06
	15	1,72	1,70	1,67	1,64	1,61	1,58	1,54
	20	2,19	2,16	2,13	2,09	2,05	2,01	1,97
	25	2,58	2,54	2,50	2,46	2,42	2,37	2,32
X −	0	2,87	2,83	2,79	2,79	2,70	2,64	2,58
	5	3,06	3,03	2,99	2,94	2,88	2,82	2,76
	10	3,16	3,12	3,08	3,03	2,97	2,91	2,85
	15	3,16	3,12	3,07	3,02	2,97	2,91	2,84
	20	3,06	3,02	2,98	2,93	2,88	2,82	2,76
	25	2,87	2,84	2,80	2,75	2,70	2,64	2,58
XI −	0	2,60	2,57	2,54	2,50	2,45	2,40	2,35
	5	2,27	2,24	2,21	2,18	2,14	2,10	2,05
	10	1,88	1,86	1,83	1,80	1,77	1,73	1,69
	15	1,45	1,43	1,41	1,39	1,36	1,33	1,30
	20	0,99	0,97	0,96	0,94	0,92	0,90	0,88
	25	0,50	0,49	0,48	0,47	0,47	0,46	0,45
XII −	0	0,00	0,00	0,00	0,00	0,00	0,00	0,00

☉'s Long.		Half Interval between the Observations.						
		H. M. II. 40	H. M. II. 50	H. M. III. 0	H. M. III.10	H. M. III.20	H. M. III.30	H. M. III.40
S.	D.	S.	S.	S.	S.	S.	S.	S.
VI +	0	0,00	0,00	0,00	0,00	0,00	0,00	0,00
	5	0,43	0,42	0,41	0,39	0,38	0,36	0,35
	10	0,85	0,83	0,81	0,78	0,75	0,72	0,69
	15	1,26	1,22	1,19	1,15	1,11	1,06	1,02
	20	1,64	1,59	1,54	1,49	1,44	1,38	1,32
	25	1,98	1,92	1,86	1,80	1,74	1,67	1,60
VII +	0	2,27	2,21	2,14	2,07	2,00	1,92	1,83
	5	2,50	2,43	2,36	2,28	2,20	2,11	2,02
	10	2,66	2,59	2,55	2,43	2,34	2,25	2,15
	15	2,75	2,68	2,60	2,51	2,42	2,32	2,22
	20	2,76	2,68	2,60	2,52	2,43	2,33	2,23
	25	2,68	2,60	2,52	2,44	2,35	2,26	2,16
VIII +	0	2,51	2,44	2,37	2,29	2,21	2,12	2,03
	5	2,25	2,19	2,12	2,05	1,98	1,90	1,82
	10	1,91	1,86	1,81	1,75	1,69	1,62	1,55
	15	1,50	1,46	1,42	1,37	1,32	1,27	1,22
	20	1,04	1,01	0,98	0,95	0,91	0,88	0,84
	25	0,53	0,51	0,50	0,48	0,47	0,45	0,43
IX −	0	0,00	0,00	0,00	0,00	0,00	0,00	0,00
	5	0,53	0,51	0,50	0,48	0,47	0,45	0,43
	10	1,04	1,01	0,98	0,95	0,91	0,88	0,84
	15	1,51	1,47	1,42	1,38	1,33	1,27	1,22
	20	1,92	1,87	1,81	1,75	1,69	1,62	1,55
	25	2,26	2,20	2,13	2,06	1,99	1,91	1,83
X −	0	2,52	2,45	2,38	2,30	2,22	2,13	2,04
	5	2,69	2,62	2,54	2,46	2,37	2,28	2,18
	10	2,78	2,70	2,62	2,53	2,44	2,34	2,24
	15	2,77	2,70	2,62	2,53	2,44	2,34	2,24
	20	2,69	2,61	2,53	2,45	2,36	2,27	2,17
	25	2,52	2,45	2,38	2,30	2,22	2,13	2,04
XI −	0	2,29	2,23	2,16	2,09	2,01	1,93	1,85
	5	2,00	1,94	1,88	1,82	1,76	1,69	1,61
	10	1,65	1,61	1,56	1,51	1,46	1,40	1,34
	15	1,27	1,24	1,20	1,16	1,12	1,08	1,03
	20	0,86	0,84	0,81	0,79	0,76	0,73	0,70
	25	0,43	0,42	0,41	0,40	0,38	0,37	0,35
XII −	0	0,00	0,00	0,00	0,00	0,00	0,00	0,00

TABLE II. *Equations to Equal Altitudes.* 111

⊙'s Long.		Half Interval between the Observations.						
		H. M. IIL50	H. M. IV. 0	H. M. IV.10	H. M. IV.20	H. M. IV.30	H. M. IV.40	H. M. IV.50
S.	D.	S.	S.	S.	S.	S.	S.	S.
VI +	0	0,00	0,00	0,00	0,00	0,00	0,00	0,00
	5	0,33	0,31	0,29	0,27	0,25	0,23	0,21
	10	0,66	0,62	0,58	0,54	0,50	0,46	0,41
	15	0,97	0,91	0,86	0,80	0,74	0,67	0,60
	20	1,26	1,19	1,12	1,04	0,96	0,87	0,78
	25	1,52	1,44	1,35	1,26	1,16	1,06	0,95
VII +	0	1,74	1,65	1,55	1,44	1,33	1,21	1,09
	5	1,92	1,82	1,71	1,59	1,47	1,34	1,20
	10	2,05	1,94	1,82	1,69	1,56	1,42	1,28
	15	2,11	2,00	1,88	1,75	1,61	1,47	1,32
	20	2,12	2,00	1,88	1,75	1,62	1,47	1,32
	25	2,06	1,95	1,83	1,70	1,57	1,43	1,28
VIII +	0	1,93	1,82	1,71	1,59	1,47	1,34	1,20
	5	1,73	1,63	1,53	1,43	1,32	1,20	1,08
	10	1,47	1,39	1,31	1,22	1,12	1,02	0,92
	15	1,16	1,09	1,02	0,96	0,88	0,80	0,72
	20	0,80	0,75	0,71	0,66	0,61	0,55	0,50
	25	0,41	0,38	0,36	0,34	0,31	0,28	0,25
IX −	0	0,00	0,00	0,00	0,00	0,00	0,00	0,00
	5	0,41	0,38	0,36	0,34	0,31	0,28	0,25
	10	0,80	0,75	0,71	0,66	0,61	0,55	0,50
	15	1,16	1,10	1,03	0,96	0,88	0,80	0,72
	20	1,47	1,39	1,31	1,22	1,13	1,03	0,92
	25	1,74	1,64	1,54	1,44	1,33	1,21	1,08
X −	0	1,94	1,83	1,72	1,60	1,48	1,35	1,21
	5	2,07	1,96	1,84	1,71	1,58	1,44	1,29
	10	2,13	2,02	1,90	1,77	1,63	1,48	1,33
	15	2,13	2,01	1,89	1,76	1,62	1,48	1,33
	20	2,06	1,95	1,83	1,71	1,58	1,44	1,29
	25	1,94	1,83	1,72	1,60	1,48	1,35	1,21
XI −	0	1,76	1,66	1,56	1,45	1,34	1,22	1,09
	5	1,53	1,45	1,36	1,27	1,17	1,07	0,96
	10	1,27	1,20	1,13	1,05	0,97	0,88	0,79
	15	0,98	0,93	0,87	0,81	0,75	0,68	0,61
	20	0,66	0,63	0,59	0,55	0,51	0,46	0,41
	25	0,33	0,32	0,30	0,28	0,26	0,23	0,21
XII −	0	0,00	0,00	0,00	0,00	0,00	0,00	0,00

⊙'s Long.	Half Interval between the Observations.						
	H. M. V. 0	H. M. V. 10	H. M. V. 20	H. M. V. 30	H. M. V. 40	H. M. V. 50	H. M. VI. 0
S.　　D.	S.	S.	S.	S.	S.	S.	S.
VI + 0	0,00	0,00	0,00	0,00	0,00	0,00	0,00
5	0,18	0,15	0,13	0,10	0,07	0,04	0,00
10	0,36	0,31	0,25	0,19	0,13	0,07	0,00
15	0,53	0,45	0,37	0,29	0,20	0,10	0,00
20	0,69	0,59	0,48	0,37	0,26	0,13	0,00
25	0,83	0,71	0,58	0,45	0,31	0,16	0,00
VII + 0	0,96	0,82	0,67	0,52	0,35	0,18	0,00
5	1,05	0,90	0,74	0,57	0,39	0,20	0,00
10	1,12	0,96	0,79	0,61	0,42	0,21	0,00
15	1,16	0,99	0,81	0,63	0,43	0,22	0,00
20	1,16	0,99	0,82	0,63	0,43	0,22	0,00
25	1,13	0,96	0,79	0,61	0,42	0,21	0,00
VIII + 0	1,05	0,90	0,74	0,57	0,39	0,20	0,00
5	0,95	0,81	0,66	0,51	0,35	0,18	0,00
10	0,81	0,69	0,57	0,44	0,30	0,15	0,00
15	0,63	0,54	0,44	0,34	0,23	0,12	0,00
20	0,44	0,37	0,31	0,24	0,16	0,08	0,00
25	0,22	0,19	0,16	0,12	0,08	0,04	0,00
IX − 0	0,00	0,00	0,00	0,00	0,00	0,00	0,00
5	0,22	0,19	0,16	0,12	0,08	0,04	0,00
10	0,44	0,37	0,31	0,24	0,16	0,08	0,00
15	0,64	0,55	0,45	0,35	0,24	0,12	0,00
20	0,81	0,69	0,57	0,44	0,30	0,15	0,00
25	0,95	0,81	0,67	0,51	0,35	0,18	0,00
X − 0	1,06	0,91	0,75	0,57	0,39	0,20	0,00
5	1,13	0,97	0,80	0,61	0,42	0,22	0,00
10	1,17	1,00	0,82	0,63	0,43	0,22	0,00
15	1,17	1,00	0,82	0,63	0,43	0,22	0,00
20	1,13	0,96	0,79	0,61	0,42	0,21	0,00
25	1,06	0,91	0,75	0,57	0,39	0,20	0,00
XI − 0	0,96	0,82	0,67	0,52	0,36	0,18	0,00
5	0,84	0,72	0,59	0,45	0,31	0,16	0,00
10	0,70	0,60	0,49	0,38	0,26	0,13	0,00
15	0,54	0,46	0,38	0,29	0,20	0,10	0,00
20	0,36	0,31	0,25	0,19	0,13	0,07	0,00
25	0,18	0,16	0,13	0,10	0,07	0,04	0,00
XII − 0	0,00	0,00	0,00	0,00	0,00	0,00	0,00

TABLE III. *For Reducing the Sun's Longitude, as given in the Nautical Almanac for Noon at Greenwich, to Noon under any other Meridian.*

Time from Noon.	Hourly Motion of the Sun.											Lon.
	2 23	2 24	2 25	2 26	2 27	2 28	2 29	2 30	2 31	2 32	2 33	
II												
0.20	0, 8	0, 8	0, 8	0, 8	0, 8	0, 8	0, 8	0, 8	0, 8	0, 8	0, 8	5
0.40	1, 6	1, 6	1, 6	1, 6	1, 6	1, 6	1, 7	1, 7	1, 7	1, 7	1, 7	10
1.00	2, 4	2, 4	2, 4	2, 4	2, 5	2, 5	2, 5	2, 5	2, 5	2, 5	2, 5	15
1.20	3, 2	3, 2	3, 2	3, 2	3, 3	3, 3	3, 3	3, 3	3, 4	3, 4	3, 4	20
1.40	4, 0	4, 0	4, 0	4, 0	4, 1	4, 1	4, 1	4, 2	4, 2	4, 2	4, 2	25
2.00	4, 8	4, 8	4, 8	4, 8	4, 9	4, 9	5, 0	5, 0	5, 0	5, 1	5, 1	30
2.20	5, 6	5, 6	5, 6	5, 6	5, 7	5, 7	5, 8	5, 8	5, 9	5, 9	5, 9	35
2.40	6, 4	6, 4	6, 4	6, 4	6, 5	6, 6	6, 6	6, 7	6, 7	6, 7	6, 8	40
3.00	7, 2	7, 2	7, 2	7, 3	7, 4	7, 4	7, 5	7, 5	7, 6	7, 6	7, 6	45
3.20	8, 0	8, 0	8, 1	8, 1	8, 2	8, 2	8, 3	8, 3	8, 4	8, 4	8, 5	50
3.40	8, 8	8, 8	8, 9	8, 9	9, 0	9, 0	9, 1	9, 2	9, 2	9, 3	9, 3	55
4.00	9, 6	9, 6	9, 7	9, 7	9, 8	9, 9	9, 9	10, 0	10, 1	10, 1	10, 2	60
4.20	10, 4	10, 4	10, 5	10, 5	10, 6	10, 7	10, 8	10, 8	10, 9	11, 0	11, 0	65
4.40	11, 2	11, 2	11, 3	11, 3	11, 4	11, 5	11, 6	11, 7	11, 7	11, 8	11, 9	70
5.00	12, 0	12, 0	12, 1	12, 1	12, 2	12, 3	12, 4	12, 5	12, 6	12, 7	12, 7	75
5.20	12, 8	12, 8	12, 9	12, 9	13, 1	13, 2	13, 2	13, 3	13, 4	13, 5	13, 6	80
5.40	13, 6	13, 6	13, 7	13, 8	13, 9	14, 0	14, 1	14, 2	14, 3	14, 4	14, 4	85
6.00	14, 3	14, 4	14, 5	14, 6	14, 7	14, 8	14, 9	15, 0	15, 1	15, 2	15, 3	90
6.20	15, 1	15, 2	15, 3	15, 4	15, 5	15, 6	15, 7	15, 8	15, 9	16, 0	16, 1	95
6.40	15, 9	16, 0	16, 1	16, 2	16, 3	16, 4	16, 5	16, 7	16, 8	16, 9	17, 0	100
7.00	16, 7	16, 8	16, 9	17, 0	17, 2	17, 3	17, 4	17, 5	17, 6	17, 7	17, 8	105
7.20	17, 5	17, 6	17, 7	17, 8	18, 0	18, 1	18, 2	18, 3	18, 5	18, 6	18, 7	110
7.40	18, 3	18, 4	18, 5	18, 6	18, 8	18, 9	19, 0	19, 2	19, 3	19, 4	19, 5	115
8.00	19, 1	19, 2	19, 3	19, 4	19, 6	19, 7	19, 8	20, 0	20, 1	20, 3	20, 4	120
8.20	19, 9	20, 0	20, 1	20, 2	20, 4	20, 5	20, 7	20, 8	21, 0	21, 1	21, 2	125
8.40	20, 7	20, 8	20, 9	21, 0	21, 2	21, 4	21, 5	21, 7	21, 8	21, 9	22, 1	130
9.00	21, 5	21, 6	21, 7	21, 8	22, 1	22, 2	22, 3	22, 5	22, 6	22, 8	22, 9	135
9.20	22, 3	22, 4	22, 5	22, 7	22, 9	23, 0	23, 2	23, 3	23, 5	23, 6	23, 8	140
9.40	23, 1	23, 2	23, 3	23, 5	23, 7	23, 8	24, 0	24, 2	24, 3	24, 5	24, 6	145
10.00	23, 9	24, 0	24, 1	24, 3	24, 5	24, 7	24, 8	25, 0	25, 2	25, 3	25, 5	150
10.20	24, 7	24, 8	24, 9	25, 1	25, 3	25, 5	25, 7	25, 8	26, 0	26, 2	26, 3	155
10.40	25, 5	25, 6	25, 8	25, 9	26, 1	26, 3	26, 5	26, 7	26, 8	27, 0	27, 2	160
11.00	26, 3	26, 4	26, 6	26, 7	27, 0	27, 1	27, 3	27, 5	27, 7	27, 9	28, 0	165
11.20	27, 1	27, 2	27, 4	27, 5	27, 8	27, 9	28, 1	28, 3	28, 5	28, 7	28, 9	170
11.40	27, 9	28, 0	28, 2	28, 3	28, 6	28, 8	29, 0	29, 2	29, 4	29, 6	29, 7	175
12.00	28, 6	28, 8	29, 0	29, 2	29, 4	29, 6	29, 8	30, 0	30, 2	30, 4	30, 6	180

TABLE IV. *Proper Stars for bringing a Transit Instrument into the Plane of the Meridian.*

Names of the Stars.	Right Asc. in Time.	An.var.in right Asc.	Declination.	An.var.in Declina.
	H. M. S.	S.	D. M. S.	S.
{ θ Eridani	2. 50. 30, 1	2, 28	41. 7. 58 S.	− 14, 75
{ β Perfei	2. 54. 53, 1	3, 85	40. 9. 14 N.	+ 14, 49
{ β Columbæ	5. 43. 44, 7	2, 10	35. 51. 53 S.	− 2, 42
{ θ Aurigæ	5. 45. 44, 1	4, 08	37. 10. 54 N.	+ 1, 25
{ α Canis Majoris	7. 15. 59, 6	2, 37	28. 54. 44 S.	+ 6, 53
{ Pollux	7. 32. 47, 1	3, 74	28. 30. 30 N.	− 7, 89
{ q Navis	10. 6. 6, 8	2, 51	41. 6. 47 S.	+ 17, 62
{ μ Ursæ Majoris	10. 10. 4, 2	3, 63	42. 31. 29 N.	− 17, 78
{ η Centauri	12. 42. 7, 7	3, 27	39. 3. 56 S.	+ 19, 71
{ 186 Ursæ Majoris	12. 46. 25, 6	2, 85	39. 25. 41 N.	− 19, 64
{ η Centauri	14. 22. 33, 0	3, 75	41. 14. 45 S.	+ 16, 29
{ γ Bootes	14. 23. 49, 0	2, 43	39. 12. 37 N.	− 16, 23
{ κ Centauri	14. 45. 53, 5	3, 84	41. 16. 5 S.	+ 15, 02
{ β Bootes	14. 54. 13, 8	2, 26	41. 12. 20 N.	− 14, 53
{ ζ Herculis	16. 33. 35, 6	2, 29	31. 58. 44 N.	− 7, 39
{ ι Scorpii	16. 36. 56, 7	3, 91	33. 54. 6 S.	+ 7, 11
{ β Lyræ	18. 42. 31, 2	2, 21	33. 8. 19 N.	+ 3, 70
{ ζ Sagittarii	18. 49. 33, 9	3, 83	30. 9. 23 S.	− 4, 30
{ γ Lyræ	18. 51. 16, 4	2, 24	32. 25. 6 N.	+ 4, 45
{ τ Sagittarii	18. 54. 7, 7	3, 76	27. 57. 5 S.	− 4, 68
{ α Microscopii	20. 37. 8, 0	3, 78	34. 31. 37 S.	− 12, 70
{ ι Cygni	20. 37. 53, 0	2, 39	33. 12. 21 N.	+ 12, 75
{ Fomalhaut	22. 46. 16, 6	3, 32	30. 42. 9 S.	− 19, 02
{ β Pegasi	22. 53. 51, 2	2, 87	26. 58. 20 N.	+ 19, 22
{ δ Sculptoris	23. 38. 13, 6	3, 14	29. 15. 46 S.	− 19, 96
{ α Andromedæ	23. 57. 48, 5	3, 06	27. 57. 41 N.	+ 20, 04

The Places of the Stars in this Table, are given for the Beginning of the Year 1795. They must be reduced to the Time they are wanted for, by multiplying the annual Variation by the Number of Years and Parts of a Year which have elapsed, and adding the Product to the Right Ascension in the Table: but it must be added to, or subtracted from the Declination, according as the Sign + or − is annexed to it.

TABLE V. *For calculating the Deviation of Stars in Right Ascension.*

Star's right Ascension	0 Signs. — 6 Signs. —		1 Signs. — 7 Signs. —		2 Signs. — 8 Signs. —		Star's right Ascension
°	° ′	Log.	° ′	Log.	° ′	Log.	°
0	0. 0	9.804	6.45	9.778	7.48	9.716	30
1	0.15	9.804	6.54	9.777	7.41	9.713	29
2	0.31	9.804	7. 3	9.775	7.32	9.711	28
3	0.46	9.804	7.12	9.773	7.23	9.709	27
4	1. 1	9.803	7.20	9.771	7.14	9.707	26
5	1.16	9.803	7.28	9.770	7. 4	9.705	25
6	1.32	9.803	7.36	9.768	6.53	9.703	24
7	1.47	9.802	7.43	9.766	6.41	9.701	23
8	2. 2	9.802	7.49	9.764	6.29	9.699	22
9	2.17	9.802	7.55	9.762	6.17	9.697	21
10	2.31	9.801	8. 1	9.760	6. 3	9.695	20
11	2.46	9.801	8. 6	9.758	5.49	9.693	19
12	3. 0	9.800	8.10	9.756	5.35	9.692	18
13	3.15	9.799	8.14	9.754	5.20	9.690	17
14	3.29	9.798	8.17	9.751	5. 4	9.689	16
15	3.43	9.797	8.20	9.749	4.48	9.687	15
16	3.57	9.797	8.22	9.747	4.31	9.686	14
17	4.11	9.796	8.24	9.745	4.14	9.684	13
18	4.24	9.795	8.25	9.743	3.56	9.683	12
19	4.37	9.794	8.25	9.740	3.38	9.682	11
20	4.50	9.792	8.25	9.738	3.20	9.681	10
21	5. 3	9.791	8.25	9.736	3. 1	9.680	9
22	5.16	9.790	8.23	9.734	2.41	9.679	8
23	5.28	9.789	8.21	9.731	2.22	9.678	7
24	5.40	9.787	8.18	9.729	2. 2	9.678	6
25	5.51	9.786	8.15	9.727	1.42	9.677	5
26	6. 3	9.784	8.11	9.724	1.22	9.677	4
27	6.14	9.783	8. 6	9.722	1. 2	9.676	3
28	6.24	9.781	8. 1	9.720	0.42	9.676	2
29	6.35	9.780	7.55	9.718	0.21	9.676	1
30	6.45	9.778	7.48	9.716	0. 0	9.676	0
	11 Signs. + 5 Signs. +		10 Signs. + 4 Signs. +		9 Signs. + 3 Signs. +